WINE SCANDAL

Also by Fritz Hallgarten

Alsace and its Winegardens
The Great Wines of Germany (together with André Simon)
German Wine Law 1969
Rhineland Wineland
The Wines and Wine Gardens of Austria
Wines of Germany, Alsace, Luxembourg and Israel in *Wines and Spirits of the World*
German Wines
Der Konflikt zwischen geographischer Herkunft angabe und Warenzeichen

Fritz Hallgarten

WINE SCANDAL

Weidenfeld and Nicolson · London

Acknowledgment

I would like to thank Ben Turner for
his editorial work on the book.

Published in Great Britain by
George Weidenfeld & Nicolson Limited
91 Clapham High Street
London SW4 7TA

ISBN 0 297 78815 9

Printed and bound in Great Britain by
Butler & Tanner Ltd
Frome and London

Contents

Preface

There is probably no more dishonesty in the wine trade than in any other, say those within the trade. They can justifiably point to other trades in every country, where there are constantly continuing cases of bribery, corruption and shoddy workmanship which are highlighted in the world's press. Indeed, in business as a whole it can be truly said that there are countless examples of false description, short measure, overcharging and corruption at every level. Even politics with its 'kitchen cabinets' and 'corridors of power' is badly tarnished, with frequent scandalous revelations in every country of the world.

So what is so special about the wine trade?

After woman, man's oldest love has undoubtedly been wine. Indeed, some even put it first and quote man's three joys as being 'wine, women and song'. The ancient Greeks created a pagan god for it – Dionysius, whom the Romans called Bacchus. In Judaism wine has a special position among other fruits and products. Special rules were laid down for its treatment and a special blessing accompanies its enjoyment: 'Blessed are you, O Lord, King of the Universe, who created the fruit of the vine.' Wine is necessary for the reception of the Sabbath. At the Sederservice (the Passover celebration) four cups of wine have to be drunk repeating the blessing each time. The eight-day-old Jewish boy gets his first taste of sweet wine on the occasion of his circumcision. In Christianity, wine has been raised to a Sacrament.

Throughout the centuries, man has forever proclaimed his praise of wine in prose, in verse and in song. In the sophisticated world of today, no other commodity has a more comprehensive vocabulary to define its intangible nuances.

There is no doubt that to interfere with his wine is to interfere with man himself. Nothing infuriates him more than to find that his wine is not what he believes it to be. Worst of all, there would seem to be more opportunities for fraudulent activities in the wine trade than in any other. Alas, one can rarely tell that one's wine is not honest until it has been drunk and not always even then. When one finds out later, it makes one feel such a fool for not knowing. One's very ego is attacked. Scandals in the wine trade strike at the very heart of man.

This book was first prepared for publication in 1976. A publisher to whom the manuscript was submitted, gave it to a friend and former client of mine when he was a director of a world-famous wine company, asking him to read it and report on its sales potential. He approached me and in a most friendly and polite manner asked me not to publish it. He felt that its revelations, even though they had already appeared in various newspapers and magazines, could damage the established wine merchant and even do harm to the wine trade as a whole. I withdrew my book but whenever I mentioned it I found an interested listener and more and more people said that this book had to be published.

The German wine scandals in ever increasing numbers and the Austrian 'anti-freeze' allegations during the summer of 1985 finally persuaded me to bring the book up to date and publish it. My aim is to alert wine merchants and true wine lovers in the consumer countries (e.g. Great Britain, Denmark) to the trickery practised by some members of the trade growers and merchants in the wine producing countries. Fortunately it is always the question of the black sheep, but an alert public will cut short their dishonest career.

1

What is a Wine Scandal?

To put it in a nutshell, any treatment, labelling or selling of wine that contravenes the law is a scandal. What is prohibited today may be allowed tomorrow, and vice versa. To give an example, in 1465 a Councillor in the town of Cologne was expelled from the Council and his licence to trade was withdrawn because he had used sulphur in the making of his wine. Sulphur has long been recognised as a powerful purifier. Sulphur candles used to be burnt inside casks to make them wholesome; indeed, they are still used in some country places. Actually *adding* sulphur to wine was another matter and created fear in the minds of sensible people. Opinion weakened somewhat over the years and small quantities of sulphur were permitted to be added, but eminent authorities were still of the opinion that too much sulphur in a wine would be dangerous to health.

Adding sulphur to wine is now regulated by law. It is usually added in the form of potassium or sodium metabisulphite, which releases the gas sulphur dioxide into the wine. This inhibits re-fermentation, prevents infection from spoilage micro-organisms and preserves the wine from oxidation. Dry wines need less sulphite than sweet wines. So, what was once a scandal is now permitted within specified limits and approved, to ensure that the wine remains in good health.

2

Some Wine Scandals Through the Ages

There is sufficient evidence from historical sources for us to be sure that ever since there was trade in wine, there has been adulteration, false description, short measure and similar scandals. English governments have been enacting decrees controlling wines, spirits and beers from earliest times. King Edgar of England (959–975) decreed a standard size for drinking vessels. How successfully his law was enforced and obeyed is uncertain, for the matter was referred to again in Magna Carta in 1215.

In 1285, Edward I regulated the price of wines in taverns and included in the Act were the words '... and if the taverners exceed, their doors shall be shut'. It is evident that some taverns were charging scandalous prices for wine, just as some hotels and restaurants do today. The wine was of a much lower alcohol content than that produced today. The vine was not so well selected, nor its cultivation so well controlled. The berries were consequently of poorer quality, producing relatively low alcohol wine that was often mixed with honey. Good ale was often as strong as the wine.

Wine goblets of silver or pewter were relatively rare in medieval England. Ordinary people drank wine from a cup or bowl made from leather or wood that was also used for ale or mead. The best of the wooden vessels were turned by carpenters on a hand lathe from bird's-eye maple and were called mazers. The name derived from the German word *mase* meaning spot – a reference to the spotted appearance of the wood. Mazers were of varying sizes depending on the part of the tree from which they were formed. Later they were embellished with patterns or carvings, then pewter and even silver bands. They were extensively used throughout the fourteenth, fifteenth and sixteenth centuries. Sometimes a person would own his own mazer, but more often it was in common use and shared by each one present, taking a sup and passing the mazer to his neighbour.

Both the price and the measure of ale and wine sold by taverners was controlled by Edward I's 1285 Act. A gallon of wine had to be sold for 'threepence and not dearer' and the only measures to be used were the gallon, the pottle and the quart. Later the pint measure was added to the list. These measured vessels were sometimes made of clay so that when they dried out they shrank and held less. Accordingly it was decreed that all vessels of measure had to be taken four times a year to the local Alderman to be checked and if found accurate to be sealed as being of the correct capacity. Such strict regulations clearly imply their need. Short measure was a common scandal and it was forbidden to sell wine or ale by 'hanaps', the cup that the taverner was obliged to supply for the use of the customer. The wine was bought by the selected measure and then poured into a hanap as required. Punishment for conviction of giving short measure was 'to play bo-peep through the pillory'.

In about 1500, the Venetian envoy sent back a report to the Doge in Venice that in England, or at any rate in and around London, 'Few people keep wine in their houses, but buy it for the most part in taverns; and when they mean to drink it a great deal, they go to the tavern and this is done not only by the men, but also by the ladies of distinction.' He went on to explain that 'at an entertainment where there is plenty of wine, they will drink ale and beer in preference to it'. This may have been due to the poor quality or high price of the wine, the high quality or low price of the ale, or to unsophisticated palates not so accustomed to wine.

Previously, in 1466, at the installation of George Nevile as Archbishop of York, the drink supplied to the guests included 100 tuns of wine and 300 tuns of ale. A tun was equal to 252 gallons. This gives us a useful guide to the proportions preferred, the number of guests and their enormous capacity for alcoholic liquor.

Wine was certainly drunk in sufficient quantities for it to be of concern to William Turner, a physician to Queen Elizabeth I. In 1568 he wrote the first book on wine ever to be printed in England. The theme of the book was that the drinking of red wine caused stones in the kidneys and bladder - a painful, serious and widespread ailment of the period. Turner advocated the drinking of Rhenish white wine that he reckoned would cool the blood and prevent the stone.

In 1660, an Act of Parliament was passed 'For the better ordering of wine by retail, and for preventing abuses in the mingling, corrupting and vitiating of wines, and for selling, limiting the prices of the same'. Heavy penalties were included for selling mixed or adulterated wine. But the swindlers were not put off and many more Acts were passed during the reign of Charles II. They indicated the considerable concern of the government at the many fraudulent and scandalous activities that were evidently widespread.

It was mostly in vain. In the eighteenth century a *Lexicon of Deceits* was published that injured many a drinker in his wallet, if not in his stomach. The author studied the adulterators closely. 'They deceived their customers by bulking out the wine with water, or good growth with bad ones. By falsifying its taste and colour with additions of sulphur, or by unslaked lime. By putting inferior musts into casks containing yeasts from high quality Spanish, Italian, Hungarian or Rhenish growths and, after fermentation, selling the wine as real. By putting fragrant herbs into the must or wine and passing it off as muscatel. By adding sugar, sultanas, syrup and a measure of Spanish, Italian or Hungarian wine to home grown wine and selling them as imports from those countries. By adulterating the products of their own cellars with sulphur, brandy and Spanish wines and passing them off as strong southern wines. By falsifying the drink with fruit juices and by using smaller measures than those prescribed in the region.'

More books were published and sold well. In 1808, John Davies, a wine merchant, published the 7th edition of his *Innkeeper and Butler's Guide*, subtitled *A directory in the making and managing of British wines*. In addition to providing recipes for making wines such as English 'Claret', English 'Champagne' and 'pearl' gooseberry wine, the Guide suggested a number of remedies for rejuvenating foreign wines that had turned sour!

In 1814, Philip Percy Carnell published a recipe for making 18 gallons of artificial claret. It included:

2 pecks of claret vine leaves
40 lb raw sugar
½ lb powdered red tartar
6 handfuls rosemary leaves
6 oranges – peel and juice only

1 gallon or more of brandy
18½ gallons of soft water

No mention is made of yeast, whether baker's or brewer's. The recipe would make a total volume of 23 gallons, so presumably as much as 5 gallons was lost in the frequent racking of the clearing wine from its lees.

Only three years later in 1817, R. Westney published *The Wine and Spirit Dealer's Vade Mecum*. It contained instructions for managing, flavouring, preserving and recovering wines and spirits, with a collection of approved recipes for making British wines.

In 1829 came another book with the intriguing title of *Wine and Spirit Adulterators Unmasked* by an anonymous author who called himself 'One of the Old School', whatever that meant. Was he a reformed adulterator of wine?

Many books appeared giving details on how to make fruit wines, meads, ciders and other drinks for use in the home. Most of the beverages were indeed consumed in the home, and were not offered for sale or passed off as anything other than a country wine. Nevertheless certain people made these wines on a commercial basis and mingled them with imported wines to improve them.

In other countries the situation was no different. In Germany, wine adulterators who were caught in 1490 were dipped in the Rhine and their casks were staved in. Practical hints on 'How to test the quality and durability of wine', were given in 1875 by Heinrich Von der Serge in a book called *Weinprobe*.

In France, the Comte de Chanteloupe, Jean Antoine Claude, published in 1801 *L'art de faire, gouverner et perfectionner les vins*. The author advanced, for the first time, the theory later to become known as 'chaptalisation' – the practice of adding sugar to the must during fermentation to improve the final product and ensure that the alcohol content was equal to the wines of other qualities.

In 1856, a book was published on how to make wines without grapes for less than three francs per hundred litres. *Vin sans raisins*, indeed! This was followed in 1882 by Joseph Audibut's work entitled, *L'art de faire les vins d'imitation*. It included recipes for Madeira, Malaga, Vermouth, bitter syrups, infusions, liqueurs, 'avec les vins de raisins, sec et autres'.

Although published in New York in 1858, a book called *Bordeaux Wine and Liquor Dealer's Guide* clearly had a French connection. The anonymous author, 'a practical liquor manufacturer' wrote what he described as 'A treatise on the manufacture and adulteration of liquors'. The market for such books was very substantial for many of them went into a number of editions.

In Italy, too, as far back as 1785 a book was published with a title that needs no translation, *Delle Adulterazioni dei vini*.

Today, the books on the management of wine are more technical and less obviously encouraging of fraudulence. Perhaps the techniques of adulteration are by now second nature!

3

How Wine Rules are 'Bent'

Such was the headline of an article by a French merchant in the trade paper *Off Licence News* dated 25 March 1976. The French merchant was an importer and seller of Italian wines. He told an informal enquiry 'I am not a fraud - no more than other merchants.' He claimed that the production conditions in French vineyards were so 'degraded' that many wines could not be sold in their original form and that this resulted in some bending of the rules in the merchants' cellars. He listed these activities as follows:

1. Increasing or decreasing the acid content of wines by illegal methods.
2. The use of sulphuric acid to rid the wine of the unpleasant taste left by some preservatives.
3. The pretentious labelling of *vins de table*.
4. The substitution of *vins de table* for AOC wines by switching certificates of authenticity.
5. The use of false Appellations, e.g. 'Such-and-such Macon' when in fact it was made with 60% Midi wine, 20% Italian and 20% Burgundy.
6. White wines coloured red by artificial means.
7. Re-naming Algerian and Italian wines as French.
8. The illegal mixing of different Appellation grapes at the harvest.

When asked by the inquiry whether the laws relating to chaptalisation (adding sugar to wine during fermentation) were not recognised in *any* vineyards he replied, 'According to what I hear, I am of that opinion - but I cannot confirm it'. He was also asked to confirm an allegation that 40% of all Appellation d'Origine Contrôlée (AOC) wines contained other wines, illegally. He declined to answer and advised the enquirer to contact the EEC lawyers in Strasbourg. He did say, however, 'If the French wine laws were applied "strictly to the letter", then 70% of all the wine now sold in France would never reach the shops.'

WHY?

The merchant's cellars had been attacked and destroyed by rioting French wine growers angered by the importation of cheap Italian wine into France. The growers contended that the Italian wine was ruining their livelihood. Aware of the problem, the French Government had already imposed some controls on the import of Italian wine. The EEC lawyers had contended that this action was illegal and they had referred the matter to the European Court of Justice. The growers, on the contrary, did not regard the controls as tough enough and cited the fact that in 1975 some 700 million litres of wine had been imported into France from Italy - almost twice the quantity imported in 1974. They threatened to pour all the Italian wine they could find into the ditches.

Full of righteous indignation, the growers destroyed tankers of Italian wine as they entered the country and attacked the cellars of those merchants known to import Italian wine, destroying their stock. Feelings ran so high that shotguns were used and two people were killed. Police had to use tear gas to disperse the crowds and maintain an unstable peace. Growers retaliated with threats to bomb the offices of the Crédit Agricôle.

The enquiry previously mentioned began in 1970 and was connected with the fake Italian wine described in Chapter 28. It was still going on in the summer of 1977 when details were again reported in *Off Licence News*.

At one time, the merchant who explained how to bend the wine rules was charged with six others for passing off as Italian a wine made from Greek and Bulgarian musts. They were all convicted and the merchant was sentenced to imprisonment for two months. He immediately appealed on the grounds that the police had got their facts wrong.

The merchant pointed out that he owned two companies. One imported and sold Italian wines, and the other was a transport company whose services were available for hire to anyone who required them.

Only his transport company had been involved in the fake

Italian wine case and then only as a carrier going about its legitimate business. The merchant argued that even if the wine was illegal, it was not the responsibility of the carrier to find out whether there was anything wrong with the goods he carried.

The truth in these matters is extremely difficult to establish beyond doubt. This case also shows how long some enquiries take. The whole situation is most complex. Apart from greed for money, there are many other factors involved, including, sometimes, even that of survival by the lowest persons involved.

In the countries that make up the European Economic Community, far more wine is produced than it can drink or sell, as much in fact as 115% of its needs. To make matters worse, EEC prices are almost double those in Spain, Algeria and Yugoslavia, so that competition in selling wine outside the Community is intense. Unfortunately, a good deal of the unwanted wine comes from low quality vines that should really be dug up. At present, however, growers of these vines are too poor to do so. Most of them own too little land to be able to improve their situation, even if they knew how. The Commission in Brussels has proposed pensions for ageing growers who leave the land, grants for grubbing up low quality vines and a complete ban on all new plantings for the next few years. The problem is not only the sheer economics of survival, but also of the peasant's pride in owning a piece of land, no matter how small. This aspect is almost insoluble.

4

My Early Years and a Few Memories

I was born in a small village called Winkel on the north bank of the Rhine. It lies within sight of the great vineyards of Schloss Johannisberg and Schloss Vollrads and is only a few kilometres from the village of Hallgarten. Like the other village children I attended the local elementary school and then the grammar school at Geisenheim where I obtained my Higher School Certificate before going to the Universities of Heidelberg and Frankfurt to study law.

As a child I was in daily contact with the villagers and often went into the vineyards surrounding the village and into the wine cellars built by the Romans long ago. I heard the daily conversation of the adults about the weather, the vines, the price of grapes and of wine. I heard, too, the gossip and rumours about this grower and that merchant. Everyone was connected with wine. Even those who worked in the towns or in factories, owned and cultivated a small vineyard. Some were even called 'Sunday Growers' because they could only tend their vines during the evenings and on Sundays. Although not descendents of serfs who knew no freedom until the Napoleonic wars, as in other regions, most of the villagers were dependent tenants of big landowners or of the church.

At a very early age I got to know the slyness of certain growers. When they brought a cart load of grapes for sale, the ox or mule would first be unharnessed so that only the wagon and its contents would be weighed. After the grapes had been emptied, the wagon was weighed again. The difference between the two weights was supposed to be the weight of the grapes for which payment would be claimed and made. Often a surreptitious foot would be seen on the scale, or a tub would be left in the wagon with the grapes, or the grapes would all be in tubs each containing two or three inches of water.

There was a rule that when carting grapes for sale, a grower was not allowed to pass his house, but must go straight from the vineyard direct to the buyer. Alas, this rule was often broken in an effort to find new ways of making the grapes appear to be heavier than they were. As soon as they were received, my father, a well established wine merchant, would check the sugar content of the grapes. Mostly the grapes from the the different growers would produce a similar result. Occasionally, however, one would be well down, indicating that the grower had fallen to temptation. He had not passed his home before making the delivery but had left some water in the container before picking his grapes. The grower never dared to contradict the evidence because the gossip that would follow would be a demoralising punishment. Thus the fear of gossip prevented many scandals.

In December 1916, my father was called up for the army. He commented: 'If the Kaiser needs me at forty-five years of age to win his war, he has already lost it'. There was no getting out of it and on 11 January 1917, my father reluctantly joined the German army. My mother and I were left alone in Winkel to continue the wine business as best we could until his return. Naturally some of the more rascally growers tried to exploit us and take advantage of a fourteen-year-old schoolboy.

MY FIRST SALE

One day an enquiry came from a merchant in Bingen for some wine. I went to the village of Hallgarten to obtain samples and took them to the merchant. He liked them and ordered a few casks. I had become a broker! I hurried back to Hallgarten and placed a firm order for the wine. Then, in the manner that was usual and which I had been taught, I drove a nail firmly into the bungs of the chosen casks and covered the head with sealing wax in which I impressed my father's seal.

It was the custom in those days to pay cash when the wine was collected, since few people had bank accounts, least of all the village growers who kept their money in some more or less safe place in their homes. On the day the wine was to be collected, the merchant from Bingen decided to accompany me to the grower, fearing to trust such a small boy with so much money.

When we arrived at the place where the wine was stored, he asked to taste the wine, a not unusual request. The grower lifted out the bung and gave the merchant a sample of the wine. The merchant immediately said, 'This is not the wine I sampled and selected. It has been watered.' The grower protested that this could not be so, since I had affixed the seal to the cask from which the first sample had been taken and, as the merchant could see, the seal was unbroken.

The merchant examined the bung and seal and observed that the bung had been so treated that the wax could not adhere to it firmly. He claimed that the seal had been lifted, the bung removed, some wine had been taken out and replaced with water. The grower denied all this. The merchant was confident, however, and said, 'I will take a sample from the cask and have the cooper seal up the cask, then I will take the sample to the public analyst and we will abide by his decision. Alternatively, I will go around the village and buy an equal quantity of wine of the quality of the original sample and you can pay the difference between what I now have to pay and what we had agreed.' Fear of a scandal, and knowing that he had been caught, caused the grower to admit his trickery. We went to a relative, bought some good wine and that was the end of the matter. By myself I would never have discovered how the bung could be removed without breaking the seal. I had learned a valuable lesson.

A FRUITFUL FRAUD

For centuries one of the greatest frauds has been the blending of cider or perry with wine. In taste, perry in particular makes the wine appear soft and fruity. The possibility of detection was minimal until it was found that the presence of sorbitol in excess of 100 mg per litre indicated such forbidden blending. Sorbitol is a chemical formed by the reduction of glucose. It is produced in very minute quantities in the grape during the ripening process, but no more than 100 mg per litre should be found in the finished wine. Any quantity in excess of this figure clearly indicates that other fruit juice, as well as grape juice, has been used in the preparation of the wine. As recorded in the *Dictionnaire du Vin* this problem has never been acute in France, although it is known in Italy and, as I know, in Germany. From the turn of this century the scientists were able to discover these illegal blends and a great number of growers who took the risk of offering such blends for sale were then found out and punished.

There was a village on the Nahe in Germany where it was discovered that there was not a single grower who had not practised this art of blending. No broker or wine merchant was willing to enter the village and buy the wine. Arrangements had to be made for the gathering of the grapes and the making and marketing of the wine to be done under supervision for a few years. Then the village was amalgamated with its neighbour and its name disappeared. A new generation has since grown up and the past is forgotten.

There was a much worse scandal some years later. A small village, that has now been merged with another and so lost its name, grew apples as well as grapes. The villagers made wine from both fruits and blended them together to make a fresh and attractive wine that had something of a local reputation. They kept their secret close to their hearts and everyone else believed that the wine was made only from grapes!

But other villages were not free of scoundrels. I remember one grower, a personal friend of my father who had a great respect for him and the wine he made. Now when I went to university we students were too poor always to buy true grape wine. Frankfurt is famous for its apple wine taverns, two of which were preferred by my fraternity. It was cheap and pleasant and enabled us to enjoy ourselves inexpensively. As a result of drinking so much of this wine, I developed a fine nose for apple wine and could detect it even when mixed with grape wine.

During a vacation, my father took me for a tasting of wines made by this friend, a leading grower and producer of fresh tasting and thirst-quenching wines. As soon as I smelled the wines, I knew that they contained apples and quietly said so to my father. My father was furious with me. He criticised my ability and declared that it was only the malic acid in the under-ripe grapes that I could smell. Although certain of my opinion, I dared say no more, although I was lectured all the way home. Long after this visit the Wine Controller called and asked if there was any of this wine left and was handed a few bottles. My suspicion was soon confirmed. A complaint had been laid against the grower. It was found that for years he had bought some special pears and apples which he added in small quantities to make the wine softer and give it a fruity bouquet. My father's friend was fined and imprisoned.

MY FIRST CASE

The first criminal case in which, as a very young lawyer, I represented the defendant, concerned a commercial traveller selling ladies' underwear. He had taken up a side line, selling Black and White Scotch whisky, to hotels in Wiesbaden, Mainz and Frankfurt. He firmly claimed that he was 'ignorant of the origin of the contents of the bottles' and sold them in all good faith. It turned out, however, that a consortium of three people regularly went round the bars of the three towns collecting the empty bottles of Black and White as they became available. They had previously made arrangements with the barmen on how to open the bottles in order to keep them in good condition for further use.

The recipe for the preparation of the 'Scotch Whisky' that was used to refill them was quite simple. There is near Wiesbaden a small town called Dauborn where a lot of corn schnapps is made. It is known in the neighbourhood as Dauborner Schnapps. The consortium bought this white spirit in bulk, coloured it with caramel, bottled it in the Scotch whisky bottles they had collected and supplied them through my client to the hotels and bars in the area. This went on for a long time and there was never a complaint as to the quality of the 'whisky'. They made plenty of profit and all concerned were very satisfied, including the management of the hotels who were getting a very good service, and their customers who made no complaints - they must have enjoyed their favourite Black and White 'Scotch whisky'.

All went well until one day the official representative of the company marketing Black and White whisky called at the hotel Nassauer Hof in Wiesbaden. He was curious to know why there were such small orders for Black and White whisky coming to him from the locality and particularly from this hotel.

The hotel manager was very surprised to hear this and showed the representative his invoices for the spirits that had been supplied only a few days previously; indeed, they still had plenty in stock. Somewhat surprised, the representative went to the bar, inspected the bottles and tasted the contents. He did not need any expert analysis to convince him of what had been done. A visit to a Wiesbaden lawyer followed and without further hesitation the Public Prosecutor was informed. The spirits still lying in the hotel were confiscated, the culprits were soon discovered and arrested and the whole story was revealed in court. It remained for me to plead for leniency. The result and verdict was three months imprisonment for offences against the trade-mark law and fraud.

One of my worst experiences in the wine trade after I had come to London was as follows. I had bought some estate bottled wines from two estates at an auction in Trier. Part of the same parcels were also sold to a wine merchant on the Mosel. I quickly sold all of my parcel of wines to my English customers and needed some more, so I wrote to my father and asked him whether he would approach the other buyer and see whether he had any of his parcel left. My father went to the merchant who looked at his stock book and said that he had not sold any of that parcel and that my father could buy it from him. My father did so and, making only a small profit for his trouble, had the consignment shipped direct to me in London by the merchant.

When I received the stock, I naturally tasted it at once and immediately refused to accept it. The cork was the same branded cork as before; the label was the same and the capsule was the same. There was even the same cask number on the label, but I was sure that the wine was different. I asked the merchant, through my father, whether he could explain why the wine was so different. The merchant replied that there must have been some mistake by his cellarmaster and the wine must have been dipped from the wrong bin. He told me that if I would keep the consignment, he would adjust the price and also supply me with the actual wine that I had ordered to prove to me that a genuine mistake had been made. The deal seemed to be a very profitable one for me and so I accepted his excuse and his offer.

The second consignment of wine duly arrived, but again (though a bit classier than the previous shipment) the wine was quite different from the first wine that I had bought and was certainly not from that original parcel. Yet the label was still the same, so the merchant must have had plenty of labels. I realised, then, what was going on and decided to have nothing further to do with this man.

It was very much later that he was found out, brought before the court and sentenced to imprisonment. He lost everything. It appeared that he simply ordered estate labels from a printer who believed that the wines were genuine and did not ask to see the certificate from the grower. The merchant also bought branded corks and capsules to match. The realisation of all the fraudulence that had been going on gave me one of the biggest shocks of my life; and I had been involved. The wines I had sold were not the same as described on their labels. It made me pause to consider whether I should even remain in the wine trade.

The greatest surprise was the selection of wines listed by the English wine trade and the taste of most of them. I was offering 1931 vintage wines. There were still many wines of the 1928 and 1929 vintages on the market and some 1921s. Up to those made in 1930, German wines could contain up to 49% of foreign wines, yet be named according to the 51% share. This was a new taste for me - sweet Hock à la Sauternes or Barsac! Leading among all other names was Liebfraumilch - Liebfraumilch Superior, Liebfraumilch Spätlese and Liebfraumilch Auslese. Some of these wines had been blended and bottled in Great Britain - all according to the German wine law of 1909. Some merchants had, before the change of law in 1930, covered their requirements for years ahead.

A PALATE FOR WINE

I remember in particular a long discussion that I had with the head of a company that was later on bought up by Booth's Distillers. The proprietor talked with me for some time about German wines and finally told me that he was surprised that someone so well informed as myself on this subject had not mentioned the best German wine from the best vineyard. Taken aback, my question as to what he meant was soon answered. He was thinking about Liebfraumilch! He was of course confused; no such vineyard existed and Liebfraumilch was but a popular generic name for a blend of different wines from the Rhine area. Furthermore, it is a much misused name. What one blender considers to be a nice agreeable wine, another might consider to be unbalanced and not harmonious. The blends of Liebfraumilch vary from brand to brand and often from year to year. You can never be certain of them. Without tasting the wine beforehand you can never be sure whether the blend is good or bad.

There is one guide, of course. Some shippers are more reliable than others. A good shipper has a nose and a palate to evaluate and appreciate a wine that another shipper may not have. One merchant confessed to me that although he always went through the motions of smelling and tasting wines when buying them, and even made notes, he had no palate, no judgement of his own. At the end of the tasting he relied on all that the salesman had told him. He said, however, that if he was ever let down by a salesman that would be the end of their association. This merchant was much respected in the trade and his constant remark was, 'As long as they play the game with me, I am a friend and a buyer, but once they let me down they will not enter my door any more.'

One of my older friends had achieved a very high reputation for his clarets. He claimed that they were absolutely genuine and excellent value for money. They were much sought after by those who believed his claim and were certainly most enjoyable on every occasion that I tasted them. In a confidential mood on a certain occasion, my friend told me the secret of his real and genuine claret. It was indeed, shipped to him direct from Bordeaux and the grapes were grown and the wine was made in the Bordelais in accordance with the appropriate Appellation d'Origine Contrôlée regulations. When the casks arrived in England there was always some natural ullage space in them, so my friend topped up each cask with two bottles of vintage character port wine or, sometimes, even well matured vintage port! As soon as the wine had homogenised and settled, it was bottled, labelled and sold to his many appreciative customers. So much for being genuine, when it was 'adulterated' albeit with a superior wine.

My friend would not agree that this was adulteration. He regarded this as the 'education' of the wine and was part of his tradition as an importer. The consumer had actually received a better wine! Here I can refer to Deinhards' History, just published. It mentions how certain Hock drinkers in Great Britain - in the middle of the last century - expected and demanded that their Old Hock had been fortified by the addition of Old Cognac.

There are still many people who just cannot leave a beverage as they find it. They think that they must improve on it, blend it and impress their own personality on it, and strangest of all, they announce proudly that they are selling an authentic article.

UNSUITABLE FOR THE TRADE

There are also small fiddlers - those who are surrounded by bottles or casks and are tempted all the time to take some for their own enjoyment. I liked my father's reserves very much and was a bad taster in so far as I swallowed every drop, thus offending against the wine merchants' code of practice. That was the reason that I should not enter the trade. My father told me repeatedly

'You will kill yourself if you swallow the wine instead of spitting it out.' So I became a lawyer, and fate provided for the wine trade to be my refuge. In a similar context there is the story of the man who was warned by his doctor, 'If you continue to drink like that you will not become old.' The man replied, 'Doctor, I know that drinking wine keeps one young, that is my reason for drinking it.'

My father's cellarmaster was very sober and tasted only what was his duty to taste. Each cask when it was laid down in the cellar, before and after racking and before buying. He tasted but did not swallow and drank only a small glass after business hours. One day in 1923 he came up from the cellar to report to my father. His voice was slurred, and he could not speak coherently. My father said, 'Basting, you are drunk.' His answer was an assurance that this was not so. 'I have not seen a drop all day long.'

A few days later when my father made a visit to the cellar at an unusual time he found a rubber tube hanging from the best cask in the cellar – a 1921 Hallgartener Wurzgarten Riesling feinste Auslese. The other end was in Basting's mouth as he sucked the nectar of the gods. My father then understood Basting's remark that he had not seen a drop all day! There is nothing more quickly intoxicating than drinking in the cellar atmosphere, laden with alcoholic vapour – especially by sucking the wine from the cask.

There is, of course, fiddling going on in the merchant's office as well as in his cellar. The boss knows and accepts this but it can become very disagreeable when the bottles are refilled with water or even left empty in the bin.

When I invited a friend to have a glass of sherry that I had bought from him, he asked me politely what I was offering him. It contained at least 30% water. My clerk denied any knowledge of the watered sherry and suggested that it was perhaps the charlady or the housekeeper. The truth came out a few days later when I returned somewhat early from my lunch. The clerk was sitting in my chair fast asleep, saliva running from his mouth and a near empty bottle of sherry beside him.

One day I received a telephone call from a friend, a large part of whose cellar was kept for clients' reserved wines, among them a large bin of Cockburn's 1908 Port kept under particular observation. There was regular stocktaking and occasionally a bottle would be missing but that did no great harm and no action was taken. Then a customer who had reserved twenty-five dozen bottles asked for a part to be delivered for a special family occasion.

The cellarmaster reported that there was nothing left in the bin. My friend went to the cellar himself and found there were in fact twenty-five dozen bottles in the bin. How could the cellarmaster make such a false report and alarm him? Then he discovered that the bottles were all there but each one was empty! The cellarmaster had for years been drinking the port but always returned the empty bottles to the bin so that to the stock-taker the bin appeared full and in order.

One of my cellarmen had a rubber hot water bottle attached to the inside leg of his trousers and for a long time he left the cellar each night well furnished for an evening of drinking. After he had been found out, the police went to his home and found labels, bottles and cases from his previous job where he had supervised the daily despatch. When the carrier called for the collection he handed him one or two cases addressed to himself and paid the carriage! Before I engaged this man I obtained a good reference from his previous employer. When I told him what had happened he was quite unaware of the traffic of his goods.

SPANISH WINE

The House of Hallgarten used to offer some very superior Spanish wines at a fair price for their quality and had something of a reputation for them. Subsequently other merchants offered wines from what appeared to be the same district, but at a much cheaper price. Hallgarten's sales declined and eventually it was decided to remove these wines from their list.

In 1970, I went to Spain to explain to our supplier the reason why we would not be ordering any more of his wines. My wife and I were actually in his office when in came an irate customer. He complained bitterly to the Spanish shipper that the wine he had just received was of such poor quality that he was contemplating suing him. The Spaniard asked him a few questions and then turned the argument round and blamed the customer – another English importer whom I knew slightly. The Spaniard claimed that it was the English chlorinated tap water that was at fault and that the importer should have used distilled water like everyone else!

I listened with amazement as the full story came out. The natural wine that I had been importing had an alcohol content of between 11.5% and 12% by volume. The same Custom's and Excise tax was payable on this as on a stronger wine of 14% alcohol. The Spanish shipper upon request and payment of an appropriate small fee, would add sufficient spirit to the wine to increase its content to the maximum permitted by the British Customs. When the wine was received in England it was diluted with water until the alcohol content was only between 10.5% and 11%. The wine could then be sold at a normal profit but at a price to undercut competitors completely as had happened to me.

In England I learned that investigations were going on against this importer, but HM Customs could not see any possibility in taking action for adulteration by water. Stranger still was the attitude of one important customer to whom we had told the reason for stopping the importation of Spanish wines. He was quite aware of what had been going on and was still prepared to buy an adulterated wine.

Upon retailing this story to other members of the wine trade, I was surprised to learn that a number of importers used to have spirit added to their wines in Spain. I understood that Port and Sherry were imported at 40° Proof (24% alcohol). They were then able to dilute the wines in England and thus evade Customs and Excise duty on a substantial part of their wine. Recently this has become more difficult if not impossible as it would offend against EEC regulations.

5

Tales from Austria

A most astonishing document was published in 1976 by the information service for consumers by the Chamber of Labour and Employees in Salzburg, Austria. The report describes 76 bottles of wine bought at random in 12 stores and from 11 wine shops and merchants on 30 December 1975. The contents of each bottle were chemically analysed by the Public Analyst and also tasted by an official panel of experts. The results made very sorry reading for everyone.

Only 23 of the 76 wines were genuine and as described by their label; 53 had some defect or other, described as follows:

4 were slightly faulty – 'etwas mangelhaft'
26 were definitely faulty – 'mangelhaft'
13 were wrongly labelled – 'falsch bezeichnet'
5 were illegal and ought not to be sold – 'nicht verkehrsfahig'
5 were adulterated – 'verfalscht'

The tasting panel described the wines thus:

'entspricht' – wine in order and corresponded with description
'fehlerhaft' – wine was faulty
'verdorben' – wine was spoiled
'mangelhaft Qualitätscharackter' – deficient in the character of a quality wine
'fremdartiger Geruch' – the bouquet had a strange smell
'nicht characteristik' – not according to type
'Mitverwendung von Wasser' – diluted with water
'Wasser nachweisbar' – addition of water proved
'Zusatz nachgemachten Wein' – addition of artificial wine
'als "auserlesene Spezialität" keine Überdurchschnitlich Qualität' – although described as specially selected quality, the quality is not above average.

The report was published with a few of the technical analyses to illustrate to the average drinker the condition of the wine offered in just one town – Salzburg. There is little doubt that this depressing picture could be reproduced in every other town not only in Austria but also elsewhere.

Apart from the actual adulteration, the scandal is that the faults had been caused by the wrong treatment of the wines or the treatment was too late or with the wrong means, i.e. carelessness or ignorance. The consumer suffers all the time. This is more evident in the next story.

A toast was given at the beginning of the New Year, 1978, by the Austrian, Dr Tauss, when he introduced the 'Austrian Wine Speciality'. He said, 'This glass contains 10% Beverage Tax, 10% Alcohol Tax, 15% Service Charge, and 18% VAT. But I hope that this will not prevent you from enjoying the wine!'

In a British restaurant he could have said, 'This wine has been imported and includes export tax, shipping and insurance charges, Customs examination fees, Excise duty at not less than 56½p per bottle, a retail mark-up, VAT at 15%, all of which has been at least doubled in this restaurant and a further 10% service charge has been added to remind you that wine is a luxury and must be appreciated to the full.'

A recent copy of the *Austrian Wine Journal* reported the following incident. Austria is famous for her 'Gespritzer'. (This is known in Germany as 'Schorlemorle' and in England as 'Wine and Soda' or 'Wine with Seltzer'.) A tanker driver who was collecting wine for his firm in Gumpoldskirchen from a nearby Co-operative noticed that whilst the wine was being pumped into the container, a second hose had been coupled up to a water tank and water was being pumped in as well. He protested at once and reported the matter to his employer.

When told that the delivery would not be accepted, the manager of the Co-op expressed surprise and said that the water had been running for only a short time and that he was quite willing to go to court because the wine that he delivered was quite up to standard.

The wine merchant was not satisfied with this attitude and went to the Austrian Wine Control. He was told that the inspectors were so busy that they could not start to make enquiries into his allegation for at least six weeks. Disgruntled with this reply the merchant then went to the appropriate official in the Ministry of Agriculture in Vienna. This Official told him that there were only eighteen inspectors to look after 60,000 viticultural establishments and that he was so harrassed and overworked that if he had known what was going on in the wine trade, he would never have entered the service. Indeed, he said, he would have become a teetotaller instead.

The manager of the Co-operative then found an excuse for his action; the water was needed to clean the tank! The wine merchant had already suffered from the incompetence of the Wine Control before. Two years earlier, they had confiscated some of his wines and released them only two days before the incident described above.

6

The Austrian 'Anti-freeze' Scandal

Austria, the carefree land of wine, women and song, has been the source of many scandals in the past, some of which are dealt with in this book, but by far the worst was the 'anti-freeze' scandal that was revealed in 1985.

Soon after the publication of my book *Austrian Wines and Wine Gardens* in 1979, I came across some Austrian wines which made me doubt their authenticity. A Beerenauslese distinguishes itself by its noble bouquet, its body, richness and natural sweetness. The wine before me had the necessary sweetness and even richness, but its bouquet and body lacked certain essential constituents. My guess was that the sweetness and richness originated from concentrated grape juice, parading as the naturally unfermented grape sugar. A Beerenauslese wine has a noticeable body as a result of the natural glycerine formed during the fermentation of a great wine. When swirling such a wine in a glass, the glycerine forms tears on the side of the glass as the wine settles down. This was missing! The Beerenauslese was certainly false.

I also saw in Germany other Austrian Spätlese and Auslese wines that had been imported in casks as fully fermented dry wine. The German importers enhanced the wine by the addition of liquid sugar and glycerine and sometimes by the addition of German grape juice so that they could be classified as German wine. There is no doubt in my mind that there was a lot of collusion between exporters and importers. The Austrian wine laws still more than the German ones, base the denomination of the Prädikat wine, Spätlese and Auslese on the way the grapes are gathered. (The German word 'lese' here means gathered). When some growers started to use mechanical harvesters, the Austrian Wine Control would not allow the Prädikat classifications, even if the must showed the minimum sugar content for an appropriate attribute. In Germany these Austrian wines automatically became

Spätlese and Auslese wines. Similarly, before 1982, Eiswein was always combined with another of the Prädikat classifications, although to fulfill the legal conditions for gathering Spätlese, Auslese and even Beerenauslese grapes were just impossible when they were frozen (see page 47). The Austrian Eiswein, however, when gathered by mechanical harvesters, could be called Eiswein and nothing else, as laid down in the Austrian wine laws.

One of my clients bought an Eiswein from a well known Austrian exporter, subject to further approval before bottling. At bottling time he found that the wine had developed quite differently from his expectations, and it did not seem to have any similarity to the sample that he had originally tasted. Accordingly he rejected the wine. This Eiswein subsequently appeared on the 'black list' of the 'glykol' wines which was at the centre of perhaps the worst scandal in the history of wine.

The European wine trade had been embarrassed at the size of the German liquid sugar scandal that had involved some 2,500 growers and had dragged on during the previous four years (see page 60). Diluting wine and sweetening it with liquid sugar is undeniably a very serious matter; but contaminating wine with a potential health hazard is obviously far, far worse. Consumers, retailers, distributors, dealers, growers, Government Ministries, wine trade organisations and the media, in a great many countries are still seriously concerned at the many different scandals that were discovered during the investigations into this giant fraud, but none more so than the risk to health.

HOW IT STARTED

The scandal began as far back as the middle 1970s, when a biochemist had an idea for making artificial wine by dissolving various chemicals in water. His recipe proved successful and he was able to sell the 'wine', but the cost of producing it was dearer than producing natural wines from grape juice. One of the chemicals that he used to provide body and sweetness in his artificial wine was diethylene glycol, a chemical used in the manufacture of the anti-freeze used in the radiator of every lorry, tractor and car in winter.

Without considering whether this substance had any harmful effect, the biochemist experimented with the enhancement of inexpensive table wine into richer and sweeter wine that commands a much higher price. He was at that time working as the Kellermeister – the person in charge of wine production – at a substantial and well known firm, where more than 100 staff were employed. Indeed, it was in the firm's laboratories that the experiments were made.

The diethylene glycol secret was sold to a number of growers and dealers who transformed their inexpensive wines into expensive wines by the simple inclusion of a few grams per litre of the chemical. It was, of course, illegal to make such an addition because this substance is not listed as a permitted additive or sweetener by the Wine Law. When dissolved in the wine, however, the diethylene glycol could not be detected by any of the analytical methods in use by the official Wine Control.

GREED, AND THEN MORE GREED

This awful scandal could have continued for years had it not been for 'the sheer greed and stupidity' – according to the Austrian Deputy Trade Commissioner in London – of one of those involved. Not content with all the extra secret profit that he was making, the person concerned tried to reclaim the Value Added Tax that he had paid for a large quantity of diethylene glycol. The local tax inspector was puzzled by such a claim for a substance that he had not heard of before and referred the matter to his superior who smelled a rat and alerted the Ministry Officials in Vienna in December 1984.

ARRESTS AT LAST

Towards the end of July 1985, the Austrian police arrested the principal suspects. Nearly five million litres of contaminated wines were seized by the various authorities. The oldest wine found to be contaminated was vintaged in 1976, the youngest in 1984, but most were from the 1982 and 1983 vintages. The wines were mostly Spätlese, Auslese and Beerenauslese, although a Trockenbeerenauslese and an Eiswein were also found. Allegedly, they all came from the Burgenland, a delightful area of East Austria surrounding the great lake known as the Neusiedlersee. This shallow lake in the centre of a plain moving out to gentle slopes, creates a humid atmosphere in which *botrytis cinerea*, the 'Noble Rot', flourishes. The area all around the lake produces in great abundance many of the higher grade wines exported from Austria.

THE GERMAN INVOLVEMENT

The Austrian wines that are exported are mainly despatched in bulk containers to dealers in Germany, who bottle and distribute the wine themselves or sell it on to other distributors who may bottle it for retailers abroad.

All the Austrian wines in Germany, whether in cask or in bottle, were placed under an interdict and forbidden to be sold. Some bottles bore labels supplied by the Austrian exporters, while others indicated the name of a German importer – many of them well known firms. Several thousand wines were involved and each one was analysed. 'Black lists' were issued by the German authorities stating the name of the grower or bottler, the name of the wine and the quantity of diethylene glycol found in it. 'White lists' were also issued giving the names of wines that had been found to be free from contamination. Some German importers had returned wines in casks to the Austrian exporters as soon as the first rumours of the scandal were heard; even so, the quantity that remained blocked amounted to more than 10,000 million litres. One German importer stated that he had one million bottles of 'White listed' Austrian wine in stock that were unsaleable. No German wine merchant or supermarket was willing to display any Austrian wine.

The Austrian exporters showed a belligerent attitude over the contaminated wines found in Germany. They contended that as the wines had been exported in casks, contamination must have taken place in Germany although the truth was quite clear and well known. In order to reject these accusations, the German authorities analysed other wines, especially the German wines stocked by the importers of Austrian wines. Some glycol was discovered in sixty-five German wines, mostly from the Rheinhesse, but also from the Palatinate, the Rheingau and from Württemburg. All these wines were withdrawn from sale and returned to the bottlers. The demand for refunds ran into millions of pounds and caused many cash-flow problems. One firm became insolvent within a few weeks and went out of business. A spokesman for the firm stated that they had sustained losses of 1.2 million pounds and had 1.6 million bottles of Austrian wine that could not be sold. He claimed that the traces of diethylene glycol had got into the German wines from Austrian wines during bottling.

All the German dealers concerned were quick to deny using the chemical and claimed that their tanks must have been contaminated by Austrian wines. If this were so, it did not say very much for the vaunted German hygiene. In fairness, the German wines were only mildly contaminated, recording 20 or 30 mg per litre compared with as much as 3 g per litre found in a Ruster Beerenauslese 1981 imported and distributed in the United Kingdom. There is only one explanation for the comparatively small percentage of glycol in the contaminated German wines - they were illegally blended with Austrian wines!

From 10% to 15% of full bodied Austrian Spätlese and Auslese wines would be sufficient to improve the German wines. It follows then that the German glycol wines, as they were called, would contain only about 15% of the original Austrian contamination. One cellarmaster admitted his sins, saying that he had done it to improve the thinner German wine and win medals in the official German competition - which he often did.

The Austrian wine propaganda office issued a full page advertisement which appeared in *Die Zeit* on 19 July 1985, as follows:

No Pardon
Against the Black Sheep
of Austrian Viticulture.

*Criminal investigations have been initiated**

**Already on 23 April 1985 the authorities took all precautions. All wines found to be adulterated were confiscated. Official Report of the Austrian Wine Control.*

52,790 Austrian growers, 1,582 Austrian wine merchants, & 52 Viticultural co-operatives produce wine under the strictest official control of quality.

Some 75% of the wine produced in Austria was not only exported to Germany but also consumed there. Austria actually informed the Ministry of Viticulture in Mainz of their suspicions that contaminated wine had been exported to Germany, but the warning should of course have gone to the German Ministry in Bonn and not to a Federal State. The first concern of the officials in Mainz was the welfare of the Rheinhessen importers. They made enquiries as to the quantity of diethylene glycol dissolved in a wine that would be dangerous to health, to see whether a special licence could be obtained for permission to distribute and sell the wine. The infamous Clause 54 of the 1971 German Wine Law authorises the Minister of the Federal State to issue such licences at his discretion.

When news of the contamination was reported in the German papers in July, indignant voices were immediately raised and it was regarded as a further scandal that the information had not been made public at an earlier date. A demand was even made for the resignation of Herr Heiner Geisler, the German Minister for Health. In August, of the three top civil servants involved, Herr Stark, the Secretary of the Rhineland Palatinate Ministry for Agriculture, Wine and Forestry, accepted responsibility for the delay in telling the public of the danger of drinking Austrian wine and resigned. Dr Hans-Bernd Ueing, the Head of the Wine Department, and his deputy, Joseph Key, were transferred to other ministries.

ACTION IN BRITAIN

The United Kingdom is one of the largest importers of German wines and, in the last few years, of an increasing quantity of Austrian wine - often found to be cheaper than the German equivalent. The Ministry of Agriculture, Forestry and Fisheries (MAFF) acted very promptly, and forbade the sale of any contaminated wine and threatened to prosecute any retailer who offered a wine for sale not having checked beforehand that it had a clean analysis. A new and extremely accurate method of analysis with a very expensive and sophisticated mass spectrometer was developed in their Norwich laboratory. From 100 mg per litre discovered by gas chromatography, the scientists were steadily able to trace first 60, then 40, then 30, then 25 mg per litre.

There was almost no reliable evidence available on the toxicosity levels of diethylene glycol and speculation ranged from no effect to instant death. One American had died after drinking 70 g of the chemical at one time because he thought it was glycerine. German sources suggested that 1 mg per litre for every kilogramme of bodyweight was a safe limit. Diethylene glycol forms oxalic acid in the human body and damages the kidneys as they try to get rid of it. It is also thought to have an effect on the nervous system.

A committee of the Wine and Spirit Association commissioned a report from the British Industrial Research Association who declared that the drinking each day of one litre of wine contaminated with 35 mg of diethylene glycol was unlikely to have any untoward effect on an eleven stone man. But Consumer Associations were worried about the long term effects.

Over 40 contaminated wines - Spätlese, Auslese, Beerenauslese and Eiswein - were found and withdrawn from sale. To the great surprise of everyone in the wine trade, some 'great names' were included in the list of suppliers. It was claimed on their behalf that wines bought from growers or merchants must have contained the glycol because they had never bought or touched the chemical themselves.

Back in Austria, some 150 people are under investigation, approximately 50 of whom are already in custody pending trial. Among them are the inventers and the leading users. They knew what they did and the world is waiting for the verdict of the Austrian court of justice.

OTHER COUNTRIES CONCERNED

As the scandal spread, so the Government officials in France, Spain, Italy, Japan and the USA started to look for contaminated wines. In Japan the sale of Austrian wines was officially prohibited. For a short while all Australian wines were also prohibited because the Japanese were simply confused by the similarity in the spelling and sound of the two names. German wines were also withdrawn from sale. A few contaminated wines were found to be on sale in the USA and France had the problem of disposing of some 80,000 litres!

ITALY JOINS THE SCANDAL

It was a great surprise to everyone to learn that some Italian red wines had also been contaminated, especially the popular Barolo Kiola of the 1974 and 1976 vintages, and the Barolo of 1975. Two Barbera wines from the 1982 vintage were also found to be contaminated. A Lambrusco Bianco and a Lambrusco Rosso as well as other wines were also found to contain diethylene glycol. Declaring themselves innocent of any contaminatory activity, those concerned claimed first that the substance must have been produced during the process of fermentation which also produces glycerine. Scientists quickly proved this to be false, and so it was next suggested that the chemical must have been in the spray which was used to protect vines from moulds. This, too, was proved not to be so. It has been pointed out that diethylene glycol softens the harsh taste of tannin and enhances the fullness, or body, of the wine. Thus the taste of a poor wine is improved. It also increases the level of dry extract in a wine that has been diluted with water.

GEOGRAPHICAL ORIGINS

As far as the geographical origin of the Austrian wines is concerned, they come from all regions but mostly from the Burgenland; many bear smaller geographical designations, either Rust, Ruster or Rust Neusiedlersee. In the labelling of Austrian wines, often a district is called after a community within the district, therefore one must differentiate between a wine from the district of Rust and an original Ruster; the latter is the wine from Rust's own vineyards. Rust Neusiedlersee means wine from all the political regions around the Neusiedlersee lake. The wine growers there now face a bleak future following cancelled orders and international distrust of their wines. Most of the contaminated wine was sold under the Rust label which covers a radius seventy miles from the village. Inhabitants of Rust have launched a campaign to restore their good name, but the struggle may well be hard and long.

This awful scandal, apart from causing untold hardship to innocent growers and merchants; the bankruptcy of large firms as well as small, and much unemployment and misery, also caused the death of the Austrian police commissioner investigating the scandal. Herr Josef Mitterer was found dead next to his own revolver. He left a note that the worries of his work were too much for him to bear. Among the papers found in his possession were indications of a Mafia-like fraud committed in Trieste. According to these papers 400,000 to 500,000 hl (66,000,000 bottles) of artificial wine are said to have been produced, later exported to Germany and labelled as Austrian wine with Prädikat.

The greedy ones will have much to answer for.

THE LIGHTER SIDE OF THE SCANDAL

The scandal was not without its humour. 'Black lists' of contaminated wine were published by the different government authorities – the Austrian list reached a few thousand wines by mid-September; the German list for German wines reached 42 and the Italian only 9. All of the contaminated wines were withdrawn from sale. The wine trade organisations began to issue 'white lists' as wines were analysed and found to be free of any contaminant. Publicity people soon advertised certain wines with slogans such as 'These wines will freeze' and 'not suitable for car radiators'. The wine festival in Wiesbaden in Germany exhibited a large poster at the entrance, 'Glycol-Free Zone'.

POSTSCRIPT

In the middle of October 1985 the first Austrian conviction was announced. Herr Otto Hotzy, aged 25, pleaded guilty to the charge of adding diethylene glycol and liquid sugar to 10,400 gallons of white wine and 1,820 gallons of red wine. Because he confessed, the fifteen-month period of imprisonment was commuted by the Court to a suspended sentence. He remains a free man unless he does it again! Will the other sixty or so who are still awaiting trial be treated as leniently?

Hubert Haimerl, aged 44, an Austrian wine merchant, has been jailed for two and a half years for lacing 200,000 litres of wine with the anti-freeze agent diethylene glycol. This is the stiffest sentence so far passed down by the Austrian courts in the wake of the anti-freeze scandal.

In Germany, the Wiesbaden Prosecutor finalised the first related case. A Mittelheim grower was charged with illegally blending a 1976 Mittelheimer Edelmann Beerenauslese with Austrian wine to refresh it and make it rounder and fuller. The Court officials have imposed a fine of DM 3,600. If the grower agrees to pay he can avoid a trial in open court!

7

Germany's Wine Laws

To understand the wine scandals in Germany it is necessary to know some principles of German wine laws. Germany has number of different wine laws since 1892; most of them were concerned with the addition of sugar, sugar solution, sterilised unfermented grape juice and the blending of German wine with wine from other countries. Because it is the most northerly wine producing country, the growers have a hard task in producing grapes with sufficient sugar and without too much acid to make a drinkable and saleable wine. These are conditions in which scandals flourish and hundreds of cases come before the German courts, although few hit the international headlines.

In July 1971 the laws were revised and codified. The old method of recognising some 30,000 different designations was reduced to 2,600, although this remains a staggeringly high number for the layman to recognise and to remember. Unfortunately the method used in the reduction of the number of vineyards represents a fiddle in itself! Usually the sites of a village are divided into three classes. In the first class are those sites with the best soil, the most sunshine hours producing the best wine and normally well known – for example Piesporter Goldtröpfchen, or Bernkasteler Doktor. One expected when weeding out the site names that three, or at least two, would be left, one for each class. Instead of this correct method, the second and third class vineyards were amalgamated with the best known sites. Some good sites were made into Grosslagen and genuine names to single sites! In this way the identity of many wines has been changed.

Germany does not base the classification of its wines on the geographic origin but on the finished product, the quality in the glass. Theoretically every single metre of a German vineyard can produce quality wine, whereas in France only approximately 20% of its vineyards are entitled to the AOC designation, which is equal to the general quality wine classification in Germany. German wines are now placed in one of three categories: Deutscher Tafelwein, Qualitätswein, and Qualitätswein mit Prädikat.

DEUTSCHER TAFELWEIN (DT)

These are everyday drinking wines comparable with the *vin de table* from France and *semplice* from Italy. Deutscher Tafelwein is a blend of local wine from one to other table wine districts of Germany, i.e. the Rhein-Mosel, Bayern, Neckar and Oberrhein, which are known as 'Weinbaugebiete'. The DT cannot bear any other geographical designation.

A new class of DT has been created, the Landwein, and fifteen Landwein regions have been designated. Only when a Federal State does not make use of the permission to produce Landwein can the geographical designation of the district be used for naming the DT which must by law contain not less than 8.5% alcohol by volume.

On the subject of the new designation - Landwein - the German Wine Institute stated: 'In essence, Landwein is a step above Deutscher Tafelwein - a wine with more body, character, higher starting must weight - but which is an uncomplicated everyday drinking wine which is typical of the region from which it is produced. It must be either a trocken (up to 9 grams per litre residual sugar) or halbtrocken (up to 18 grams per litre residual sugar) wine.'

The only true fact in this description is the higher starting gravity, namely of $\frac{1}{2}^{\circ}$ of alcohol. It remains part of the Tafelwein group and contains no more body. Furthermore, it must come entirely from the region which is indicated. All the Deutscher Tafelwein have been enriched before or during fermentation, in the same way as the wines in the next class.

QUALITÄTSWEIN BESTIMMTER ANBAUGEBIETE (QbA) (QUALITY WINES OF SPECIFIED REGIONS)

Use of Sugar

Sugar to be used for enrichment must be technically clean, non-coloured sucrose and must contain at least 99.5% fermentable sugar. Sugar containing starch is no longer permitted, although it was formerly used by growers to give their wine more body and make it taste rounder. This is one of those laws which is difficult to understand. In most industries great efforts are made to get the best out of the raw material available. In the German wine industry, however, the use of sugar containing starch is prohibited. The starch is not harmful in any way; on the contrary, it produces a better wine. The reason behind the law is supposed to be the protection of the consumer who might think that such a wine was a better product and be induced by the merchant to pay more for it.

The use of syrup is not permitted. The sugar must be dissolved in the grower's or merchant's own cellar. He must not buy dissolved sugar as this could also contain other ingredients not allowed.

Qualitätswein may also be enriched with sugar before or during fermentation to increase the alcohol content to the minimum required by the law, but it has to pass several other tests for quality before a certificate of authenticity and quality can be granted to its producer. Here is the foundation for many fiddles and many scandals which have occupied the German law courts.

Geographical description

Based on the EEC regulations, the wine must originate in one of the following eleven regions: Ahr, Baden, Franken, Hessisch Bergstrasse, Mittelrhein, Mosel-Saar-Ruwer, Nahe, Rheingau, Rheinhessen, Rheinpfalz and Württemberg.

There is an enormous difference between the sizes of the various regions. The former regions were made into districts. Wine from all Baden districts between the Bodensee and Main are being blended and remain Qualitätswein, but if wine from Bingen (Rheinhessen) is blended with wine from Bingerbrück (Nahe) the result is not a Qualitätswein. This is the greatest nonsense in legislation that was ever produced since these two vineyards are only separated by the width of the River Nahe. The distance from the Bodensee to Main is some 200 kilometres!

The eleven regions are further sub-divided into districts called *Bereich* which often take the name of the best village in its area. For example, the village of Bernkastel surrounded by its famous vineyard has had to give its hallowed name to nearly 9,000 hectares of the Mosel-Saar-Ruwer region. The name Bernkastel may now be printed on the labels stuck on all the bottles of Qualitätswein produced in this large district, provided that the word 'Bereich' precedes it. Thus a Bereich Bernkastel Qualitätswein is quite a different wine, albeit less expensive, than a true wine from Bernkastel itself. The Bereich Bernkastel, however, is no more than a blend of different wines from anywhere in the district, that have reached the necessary minimum standards to be entitled to the Qualitätswein classification.

After the Bereich comes the village or commune, which in the EEC legislation is called 'local administrative area'. Village names on their own are very seldom seen because the new legislation makes it easier to use the name of the *Lage* (site) described in the EEC regulations as a 'small locality'. The viticultural area in each village is sub-divided into vineyards and the vineyards have registered names such as Marcobrunn, Doktor, Jesuitengarten etc. The combination of the name of the village and the name of the vineyard is the Lage name. The village name as a rule precedes the vineyard name in adjectival form. If the vineyard is situated in one or more villages in one piece of land as for example Erbacher Marcobrunn, it is described as an *Einzellage*, a single site. This once meant an individual vineyard situated in one village. A vineyard covering areas of two villages was, until 1971, a generic name which could be used for wines produced in a radius of 15 kilometres. But even single sites may now cover the vineyards of a number of villages and the grower has the choice of name of the precise village that he wants to append to his wine. A grower from the unknown Andel will therefore market his wine under the famous Bernkasteler Kurfürstlay label.

Many vineyards appear more than once in the vineyard register: Goldberg 27 times, Schlossberg 79 times, Kirchberg 37 times; but if such a name appears on its own it is a brand name. It denotes a geographical designation, a geographical unit, only when it is accompanied by the name of the village – as is especially laid down in the wine laws. Importers indicate brands by printing the name between inverted commas and adding the words 'Registered Brand' or ®. If used as a brand it must be printed clearly, separated from the geographic designation.

Grosslage is the name of the next geographical category. A Grosslage is a collection of a number of Lage including the name of the vineyard, that are supposed to share the same general micro-climate and soil situated in one or more villages. For example, in the Bereich Bernkastel there are a number of Grosslagen, the best known of which is Bernkasteler Kurfürstlay. A wine labelled Bernkasteler Kurfürstlay is a blend of wines from any of the fifteen different villages within the Grosslage. Although the name Kurfürstlay implies a certain superiority, the villages are in fact, not among the very best in the Bereich. Similarly, the Grosslage 'Kloster Liebfrauenberg' within the Bereich 'Südliche Weinstrasse', covers 1,340 hectares (3,310 acres) and contains seventeen villages, the leading one being Bad Bergzabern. Any Qualitätswein from this Grosslage may be a wine from any of the seventeen villages or a blend from all of them, but it still has the right to be called a 'Bad Bergzaberner Kloster Liebfrauenberg'. There is a choice of wine between the seventeen villages but the name remains the same! The 'Bernkasteler Kurfürstlay' need not contain one drop of Bernkastel wine, but a bottle labelled Bernkasteler must contain 75% of wine from Bernkastel.

A particular problem is the impossibility of telling which name is a Grosslage and which is an Einzellage unless one has a highly specific knowledge; and there are 130 Grosslage and 2,600 Einzellage names to learn!

All the different points mentioned have been the basis of wine scandals in the past, but they are nothing compared with those arising from the Prädikat names which quality wines can bear if certain legal conditions are fulfilled.

QUALITÄTSWEIN MIT PRÄDIKAT

In addition to conforming with the same regulations concerning the production of Qualitätswein, a Prädikatwein must possess additional attributes of quality. No wines in this category may have sugar added to them. The grapes must be sweet enough to make, after some manipulations, a naturally good wine. The actual alcohol content of the finished wines need be only 7%.

Kabinett

This name, once the pride of the grower for the best casks in each category - Spätlese Kabinett, Auslese Kabinett, Trockenbeerenauslese Kabinett - has become the attribute for the lowest grade of natural wine, the former ordinary *Naturweine*. They are now no more than thirst-quenchers!

Spätlese

A wine made from fully ripe grapes that have been gathered later than the lower categories. Even so, they sometimes contain too much acidity - a sign that not all the grapes were fully ripe, and the juice must therefore be deacidified. Not all of the wines legally defined as Spätlese live up to expectations of quality.

Auslese

A superior wine made from individually selected best bunches of grapes, after cutting out the bad, sick and damaged berries.

Beerenauslese

An even more superior wine made from the best individual berries from the best bunches of grapes. Only berries with 'noble rot' or at least over-ripe berries must be used.

Trockenbeerenauslese

A rare, very expensive but superb wine, made from completely shrunk berries with 'Noble Rot'. Should this not be present, because of the characteristics of the vine or because of particular weather conditions, over-ripeness of the grapes is sufficient. This last admittance of over-ripe grapes is a watering down of the traditional Trockenbeerenauslese. The grapes are left on the vine until they become dried (*Trocken*) rather like raisins and are very sweet.

Eiswein

This wine is even more rare and expensive. It is made from grapes that are frozen hard by an early frost at the time of harvesting, with the minimum sugar content of a Beerenauslese and pressed in the frozen state. At the time of writing, the Minister has indicated that the Eisweins sold in the past have spoiled the good name of Eiswein and that what I have indicated as the requirement of Eiswein has become law.

The best German winegrowers are enthusiasts concerned to produce the very best possible wines from their grapes. The *law* consequently places its emphasis on quality rather than quantity.

8

Deutschland Über Alles

In Germany, by the early 1960s there were so many criminal cases due to offences against the wine laws in the Federal State Rheinland-Pfalz, that a separate office had to be formed to deal with them under a separate Public Prosecutor. In Germany the office of Public Prosecutor is not independent of influences from Government Ministries. It is, therefore, extremely difficult to lay blame with certainty. An individual Public Prosecutor may be at fault personally or he may be acting in accordance with instructions given to him by a higher authority. Public Prosecutor Oberstaatsanwalt Bohr, achieved quite a reputation. In 1964, Bohr wrote a lengthy article about the types of criminal in the wine trade. He started by saying: 'The Rhine wine is not always pure; many German wines contain more water than permitted and wine growers and merchants adulterate and mix wines to such an extent, that the criminality is quite significant.'

He then went on to give details of the tricks they got up to. He wrote: 'Many of them are the "Great Ones" able to pull many strings to obtain a decision favourable to them. These "honourable merchants" have intimate connections with all the highly placed people in the economic and political life of the community. When necessary such people are pursuaded to put pressure on the Public Prosecutor, who is but an official who must look after his career so that he can progress to a higher income. Such a criminal plays his many cards in turn, his money, his position in society, his membership of Associations and Official organisations, such as the Chamber of Commerce and Chamber of Agriculture, and his many aspects of public life.'

Bohr goes on: 'The reason why he commits crimes is because he has substantial means and needs even more to extend and strengthen his social position and his influence in commercial, cultural and political affairs. He wishes to be considered even more important, and to appear richer and greater. The "Great Ones" rely on their connections with the Burgermeister or with the local MP who, in his mind, just cannot let him down and will surely help him. After all, he is the employer of so many people, the man who pays high rates and taxes. Since he is often a benefactor of the church, even the local priest may speak for him. He relies on assistance from all sides in his defence against the Public Prosecutor.'

Unfortunately, in many cases these so-called 'Great Ones' succeed. According to German law, many cases can be settled by *submissions*. A sum of money is negotiated, is paid and nothing further is heard of the misdeeds. Only a few cases are heard in open court, when the law will not allow a settlement by submission, or when the 'Great One' is not so great, or when he cannot pay the amount required to free himself from the burden of a criminal case.

But Herr Bohr was not immune from the influence of some 'Great One', perhaps a Minister in the Government or a very high official. For Bohr, too, succumbed and eventually was reprimanded by his superior, the Generalstaatsanwalt, who agreed that:

1. The State Domain wines in Mainz had been wrongly labelled over many years.
2. The Minister of Viticulture in Mainz and the many others involved could not be prosecuted because the wrong labelling had been authorised by Herr Oberstaatsanwalt Bohr.

The State Domain of Rheinland-Pfalz in Mainz became a member of the Central Wine Co-operative who made and marketed the wines. They used labels, however, that were exactly the same as when the Domain was independent and implied that the wines were produced by the State Domain.

After prolonged correspondence I managed to convince the Generalstaatsanwalt in Koblenz of the offences of all concerned. He agreed that I was correct and instructed the owners of the wine, the Central Wine Co-operative, to stop the use of the labelling immediately.

Strange as it may seem, I had had a long exchange of correspondence with Bohr about the labelling of the wines of the Rheinhessen Domain although I was not involved in this particular case. To my great surprise, however, I received a letter from the Generalstaatsanwalt a few weeks later, informing me that on account of new wine laws, the use of the labelling was in order and he had, therefore, withdrawn his instruction. In other words, the wrongly labelled wines could be sold and those unlabelled could now be labelled and sold by the Co-operative as State Domain wines.

The law he quoted was not a new one; it had been in force when he made out his first order. The law itself was quite clear. The labelling of the wine had to show the Co-operative as producer and bottler, but the name of the grape grower could be mentioned on the Co-operative's label. The labelling was clearly a contravention of the law, but all objections were of no avail.

The membership of the Domain came to an end in summer 1976 and the Domain made a new agreement transferring the Co-operative wine of 1974 and 1975 vintages to a new distributor. This new distributor used the same labelling - the principal label remained the Domain label and the Co-operative neck label was replaced by the neck label of the new distributor. When I found out that the labelling scandal had not stopped, I approached the Committee for the examination of quality wines to enquire what had been going on between the various offices - the Ministry of Justice, Public Prosecutors' offices in Koblenz and Mainz, the Ministry of Viticulture in Mainz and so on. I was told that 'In pursuance of Clause 54 of the German Wine Law, the Minister of Viticulture had used his discretion and given permission for the sale of the wine.'

A DANGEROUS PRACTICE

A special licence permitting distribution or export can be obtained, even for products which have been confiscated or are prohibited from sale because they are contrary to regulations, *if* it can be shown that this would avoid undue hardship and provided the deviation is small and is unobjectionable from health reasons.

This Clause 54 of the Wine Law has been stretched as though it were a piece of elastic and used, rightly or wrongly, to befriend political supporters and people with influence. For example, I was present at a meeting of the Rhine and Mosel Shippers' Association, when the Chairman expressed very special thanks to one of the senior officials at the Ministry of Viticulture at Mainz, for having shown so much understanding of the problems of the wine trade by his reasonable interpretation of the German and EEC regulations!

Actually, false labelling is one of the cases for which the special licence is *not* permitted - after all, the wine could be correctly relabelled. In spite of this the Rheinhessen District Co-operative sold many vintages of the Rheinland-Pfalz Domain that had been vinified, treated and produced by the Co-operative from grapes of the State Domain and sold with the Domain label. When the misleading labelling was discovered by the wine trade, the illegal labelling continued - licensed by the Ministry of Viticulture!

Wines rejected by the Wine Control come onto the market at cut prices that harm the legitimate trade. The German wine trade paper *Weinwirtschaft* reports that the District Government in Trier allowed the sale of two consignments of Germanised wines in the Mosel region! It was surprising that the Mosel growers did not start a riot - they only complained!

The German and Austrian authorities may promise more and better control. The controllers, however, usually living near the criminals and often entertained by them, are sometimes not reliable. In one case I had to deal with, a controller had warned the grower that he would have to visit him in two or three days time and he hoped that the bookkeeping was in order. When he arrived the books were nearly up to date but he expressed some doubts about certain wines he had tasted. During the following discussion when the merchant made the usual excuses of lack of time, shortage of staff and so on, the controller advised the merchant to pay a retainer to him and take his advice! This was not an isolated case.

WINE AND WATER

One of the worst ever scandals in Germany came before the Criminal Court in Mainz in November 1976. In the dock was the former Technical Director of a Sparkling Wine Manufacturer, accused of having diluted foreign wines with a million litres of water before converting them into sparkling wine.

The local wine controller discovered in a duty-free warehouse, containers that had arrived about a month ago. They contained 526,449 litres of wine with an alcoholic content of 12.5%, and 200,000 litres of water. The water had been pumped into the containers and pushed the very dry wine, with a specific gravity lower than that of water, to the top. The water remained at the bottom, separate and unmixed, except for the line where the two liquids actually touched.

The water was discovered by one of the workmen emptying a container. Becoming thirsty after a while, he helped himself to a glass of what he thought was wine, but promptly spat it out because it tasted so foul. Disgusted and surprised, he turned off the tap and reported the matter to the wine controller, who quickly confirmed that the 'wine' was water. The investigations started at once but took over five years to complete.

The ingredients used to make the sparkling wine, consisted of Bulgarian wine, added alcohol, chemical flavouring and preservatives, in addition to the water. Between March and August 1970 when the fraud was discovered, some seven million litres of the mixture had been turned into 'Sparkling Hock'. With one exception, all those concerned lived outside Germany and could not be prosecuted. What is significant is that the investigations started in August 1970, but the case did not come to Court until November 1976, and the hearing took nearly one year. The verdict was given in September, 1977 as follows:

1. For infringement of the Wine Law and infidelity against his employer - one year's imprisonment. But if the guilty manager paid 10,000 DM to the Staatskasse (Treasury), the sentence would be suspended, provided that no further crimes were committed in the next three and a half years.
2. For his frauds to the value of 101,000 DM - a fine of 9,000 DM.
3. The watered wine, worth 51,232 DM for distallation purposes, to be confiscated.

A Well Sugared Wine

More recently there was another case involving water. The Criminal Court in Winsberg sent a grower to prison for several years in 1984, after a hearing lasting 81 days and involving 28 experts and 34 witnesses. The grower had increased the volume of all his wine by 10% by the addition of water from his own garden well!

This reminds me of a well known story from the Rhineland about my father's cousin, Leo - a great wine expert. In a case before the Wiesbaden Court he expressed his conviction that the wine had been sugared, but the chemist who had analysed the wine denied that this had happened. Finally the wine grower confessed that he *had* added sugar to the wine, and was found guilty. The chemist then wanted to know how Leo was able to detect the sugar which was considered to be undetectable. Nobody else but the family ever got to know the answer. It was this: the grower had used water from his own well for dissolving the sugar and the sugar solution had imparted the distinctive taste of the well water to the wine. Before Leo tasted the wine he had rinsed his mouth with the well water and could taste it in the wine. He was therefore able to give his correct opinion with a clear conscience.

On further analysis of this well water it was found to contain eight times as much nitrate as is considered safe. It is all the more surprising that the grower had found a buyer for his wine.

The Sartorius Case

Weinwirtschaft called the Mainz case of 1976–7 the worst case of wine adulteration in German wine history. By making this statement, however, they forgot the case of Sartorius, in which the predecessor of the *Weinwirtschaft* the *Weinblatt*, and especially the grandfather of the present principal shareholders, took a very active part in delivering the guilty man to justice. He also had the special merit of providing stenographers who took down every word spoken during the hearing and then published the full account, thus enabling us to follow the details of proceedings taken sixty years before.

The case of Sartorius was very often quoted to me by Alsatian growers who blamed the bad state of Alsatian viticulture after the First World War on Sartorius. This falsifier of wines who undermined the Alsatian market with his cheap offers, was not just an everyday wine merchant. He was the MP of the German Reichstag who guided the German wine laws through Parliament. He approved personally of all the measures that the Reichstag enacted for the treatment of wine but objected very strongly to the keeping of records concerning wine transactions for public examination. In the end he was caught by his own private bookkeeping in a pocket notebook which showed all his sins, although some could never be found out since nobody was able to break the code he had used.

INACCURATE DESCRIPTION

A wine said to be of a specific year may legally contain 5% (formerly 25%) of wines from other years. This helps the growers and merchants to even out the bad years with the good. Poorer wines from those bad years may legally be blended with the better wines from the good years, and this may not be regarded as unreasonable by the charitably minded. But the wine is *not described as a blend* of different years and the grower or merchant may legally describe the wine as being from the good year. It is the consumer who is not getting what he thought he was paying for.

Furthermore, a wine described on the label as being made from the Riesling grape, may legally contain up to 15% (formerly 25%) of wines from other grapes. More often the grapes are mixed together at the outset, but the consumer has no means of knowing that a wine described as a Riesling is only 85% Riesling and 15% any other variety, usually producing more juice of a poorer quality. In bad and mediocre years, the unripe Riesling contains a high acidity and little aroma. This is given later to the wine by the addition of approximately 10% of unfermented but sterilised juice of grapes with natural bouquet such as the Muller-Thurgau, Morio-Muscat and Scheurebe.

Yet again, a wine said to be from a specific village may legally contain 15% of blended wine and 10% *süssreserve* from some distant part of the region. Fond thoughts that one is drinking wine exclusively from a village known to you are not wholly true - only 75% true. A 1976 Bernkasteler Doktor Riesling may contain up to 25% of a blend of wines, including *süssreserve*, from various vintages and sites from the Luxembourg frontier to Koblenz.

The German Wine Propaganda Office claims that the German wine laws are the most precise in Europe and have created 'clarity and verity'. Alas, only 75% of a German wine's description is truth; the remaining 25% is in considerable doubt. The best bet is the shipper's honesty!

In September 1978 the Juristentag had their convention in Wiesbaden. The assembly included judges, officials of the administration and lawyers. These occasions are customarily used to show the visitors something of the surounding countryside and part of the entertainment was to visit leading estates and to taste their wines.

The Stabilisierungsfonds is also much involved in litigation and this great assembly was to be used, not only to sell German wine, but also the opinion of the organisers about some pending questions of the wine laws. It was regarded as a good occasion to teach the initiated judges who may have to make decisions in wine law cases. The presentation of the wine samples and the lectures about wine law were very cunningly interconnected by the organisers. Time was allowed for questions but instead of a question I criticised the tasting notes and the interpretation of the law. I began, 'We have been told a lot about verity and clarity, but I would have preferred to have been told more about the verity and here are my observations....

(The following section is a summary of my observations on this subject.)

CERTIFICATES OF AUTHENTICITY

The other major problem concerns certificates of authenticity, as has already been mentioned. Dr Michel, the Director of the German Wine Propaganda Office has proudly declared that 'Each wine has to be officially approved with a certification number as proof of its quality.' But what does a certification number actually prove? Only that a sample submitted to a committee has been passed as having a minimum quality as prescribed by the Federal Government. What Dr Michel did not say, as a recent scandal clearly shows, was that a grower or merchant can print as many labels as he likes bearing the certification number awarded to him - there is virtually no control. This is exemplified by the prosecution in 1975 of two exporters. One had imported Italian grape juice, fermented it in Germany and then exported it to the USA as Liebfraumilch - a German Qualitätswein. The other had falsified certificates of origin and exported wines that had never been approved.

Late in 1976, the Trier Chamber of Commerce took action against a firm of wine merchants who had been expelled from the Mosel-Saar-Ruwer Wine Trade Association. The Public Prosecutor declared that false examination numbers had been used by them in many variations. In particular he claimed that:

1. One and the same wine was sold under different names and different examination numbers.
2. Different wines were sold under identical examination numbers.

3. Wines that had not received any examination number because they had not reached the minimum quality, were sold under invented numbers.
4. Even Austrian wine that had been rejected by the Wine Control as not permissible for import, was sold as Austrian Qualitätswein mit Prädikat.
5. Kabinett wines were sold under a Spätlese label.
6. Wines that had received an examination number were treated and 'substantially changed'.
7. Some chemicals forbidden by law had been used as preservatives.

The Wine Books of the firm were in perfect order and must have been manipulated to hide the real truth. The German wine trade had long been asking for this firm to be investigated because they were selling Spätlese and Auslese wines at prices that could not have covered the costs of authentic wines.

The question was also asked whether the wine managers of the supermarkets would be prosecuted as accessories. They had placed orders with the wine merchant for the delivery of wines as required by them up to 30 June 1977. Yet they must have known from the prices charged that the contents of the bottles could not have been the same as their labels described. Many people thought that the supermarket managers were equally guilty with the merchant.

Sadly, the thorough investigators of every case take anything up to five years to obtain sufficient evidence to bring a case to court and the proceedings may then drag on for many months. In the meantime, a swindler could be making plans to start another company in another place, under a different name, in the sure knowledge that nothing will happen for years. Furthermore, when a case does come before the court, the loss of memory among many witnesses about exactly what happened so many years before, makes it extremely unlikely that a hard sentence can be expected. In fact, most punishments consist of fines that are paid out of the excessive profits. The real danger to the criminal is, therefore, minimal.

9

The Grüne Woche Expo

The 'Internationale Grüne Woche' (Green Week) in Berlin is a great propaganda shopping exhibition and centre for German wines. In 1977, hundreds of older vintages were shown by the 49 exhibitors as well as 147 wines from the 1976 vintage. These new wines were made the subject of a special investigation by the Wine Controller of Berlin who reported the results in the trade paper *Weinwirtschaft*.

	Wines examined	*Rejected*
Qualitätswein	11	1
Qualitätswein mit Prädikat:		
Kabinett	29	4
Spätlese	51	10
Auslese	47	15
Beerenauslese	9	9

The Wine Controller commented, 'This list contains three Auslese wines twice. They were shown by different Estates and rejected.' The same wine produced by two different Estates? Actually, one grower used the different names of his subsidiary firms and showed identical wines. Other identical wines were shown under different labels of origin! Here are some further details:

1. One Auslese had been rejected by the Examination Committee before the exhibition. In spite of this the wine was exhibited as an Auslese with an examination number.
2. Another Auslese was labelled with quite a different site name than that stated on the application for the certificate of authenticity.
3. Twenty-eight wines (1 Qualitätswein, 4 Kabinett, 10 Spätlese, and 13 Auslese) showed certificate of examination numbers, although an application had not been placed, let alone granted.

Further investigations after the exhibition created great doubts about the identity of a wine that had received a certificate and the wine that was actually supplied to a buyer. Another wine shown as a Spätlese was later rejected as such and classified only as a Kabinett. Yet another wine for which an application for a certificate of authenticity had been made as a red wine, had been sold during the exhibition as a rosé wine!

Bestowing different names on the wines at different times has been observed again and again. The report mentions twelve cases where the application for the certificate of authenticity originated from a firm other than the exhibitor. In other words, one grower sold to another - perhaps a subsidiary - wines with certificate numbers, although no certificates had been granted. The report summed up: 'Nine exhibitors are to blame for these offences. The Authorities are making further investigations.'

Looking back on past experience in such matters, we shall not hear any more about these incidents. The guilty ones will negotiate with the prosecuting authorities and pay some nominal sum for their misdemeanours and this will be accepted by them as part of the risk. But the customers who visited the exhibition walked from stand to stand, tasted and bought the wines, and paid for them in the belief that they had bought genuine wines direct from the grower; what of them?

Well, they at least got a warning from the authorities just prior to the 1978 Green Week. The advice read as follows: 'Try to resist the offers of German wine bargains and sekt, for no fewer than 52 firms will offer you 1,200 different wines.' Further good advice was given by an official of the Institute of Food Chemistry. He told those proposing to visit the exhibition: 'After tasting a few samples and emptying a few glasses, take the precaution of immediately buying a bottle of any wine that you order in quantity. Take the bottle home with you and do not open it until you have received the case that you ordered. Then compare at once the wine in the bottle that you brought home with you, with the wine just delivered. Every year without fail after the end of the Green Week, many consumers complain to the Institute of Food Chemistry that the wine delivered against an order placed at the Exhibition was quite different in taste from the wine they tasted at the Exhibition and which they ordered.'

This is good advice for all such purchases wherever they are made. If a wine merchant unknown to you tempts you to buy and allows you to taste a wine for later delivery, always ask him to leave you one bottle of the wine for later comparison. Know your wine merchant. If you know him you can trust him. If you have not seen him before, you may never see him again.

10

Sugar in Wine

Sugar is added to wine in many regions, especially after a bad summer when there has not been sufficient sunshine to ripen the grapes. During fermentation, the grape sugars (fructose and glucose) are converted into alcohol and the gas carbon dioxide. The alcohol remains in the wine and the gas escapes into the atmosphere. In less sunny summers and even in good summers on secondary sites, the grapes contain insufficient sugars to make a wine with an adequate alcohol content. In these circumstances, wine makers in certain specified areas are allowed to add sugar before or during fermentation to increase the alcohol content of the wine to the amount required or the maximum level which the law permits. The better quality wines are made from grapes grown in more favoured areas where the situation of the vineyard is such as to catch the maximum amount of sunshine and reflected warmth. Consequently they do not need the addition of sugar, unlike wines from other vineyards in less favoured locations.

GERMANY

In the German vineyards, the most northern in Europe, only one vintage in ten can normally be rated as an excellent vintage. Another one or two may be rated as good but even in these years some German wines need enrichment to make them drinkable. Enrichment is the name given to the addition of sugar during fermentation to increase the alcohol content. More sugar is added to wine in Germany than in any other country. In the lower quality wine it is entirely legal. This has not always been so, however, and the addition of sugar or too much sugar to wine has been the cause of many scandals in the past.

German Tafelwein is enriched by sugar and/or concentrated

grape juice; Qualitätswein with dry sugar only during fermentation to ensure that the alcoholic strength reaches the minimum allowed by law. When fermentation has finished, the residual sugar content is less than 1% and the wine is so dry that almost every wine drinker would refuse it, especially when it has acidity. Indeed, in such circumstances its best use be as a component of a French salad dressing!

There is, however, a means of making the wine drinkable and enjoyable, that is by the addition of 'süssreserve'. Before bottling, the wine is fined and then sweetened with sterilised, unfermented grape juice, sometimes to 10% of the total, and bottled through sterilising filters. Some bottlers pasteurise the wine by heating it for a few minutes at a temperature of approximately 60°C before corking the bottles. The German Wine Law of 1971 permits the addition of süssreserve to all German wines, but it must be of the same category as the wine being sweetened – Spätlese süssreserve for Spätlese wine, for example.

There are different gradings relating to the residual sweetness of Tafel and Qualitätswein. Wines for diabetics may contain no more than 4 g of sugar per litre. A dry wine (trocken/sec) may contain only 9 g per litre. A medium dry wine (halb-trocken/demi-sec) only 18 g per litre. A Qualitätswein, a Kabinett, a Spätlese, an Auslese, or a Beerenauslese may contain any quantity of residual sugar as long as the alcohol content is at least 7%.

According to my own interpretation of the German law, Auslese and especially a Beerenauslese should not be made by the addition of süssreserve. The residual unfermented sugar in these wines should be the natural sugar of the grape, unfermented because of its high concentration and the 'Noble Rot'. Beerenauslese with grape juice added is, in my opinion, undrinkable. Yet there are some growers who pick grapes attacked by the 'Noble Rot' and leave them unfermented to what is in my opinion, a faked Beerenauslese. German wine trade papers contain advertisements offering süssreserve of Auslese and Beerenauslese quality! Süssreserve does not count as blending. Its addition counts as treatment and opens the way for fraud.

The lowest quality German Prädikat wine is not good because the minimum sugar content of the Prädikat Kabinett, Spätlese and Auslese, as laid down in the wine laws is too low. There is for the sake of argument a space of approximately 10° Oechsle (specific gravity) between each category. The Kabinett (76–85° Oechsle) which is made from the riper and therefore sweeter grapes (85° Oechsle) will produce a better final product and will show after the added grape juice more than the minimum alcohol content of 7%.

This is where those who select and supply become important. The consumer can choose whether to buy from an expert who carefully selects the wine, or from a merchant who buys the label of the cheapest possible price, and therefore always the wine of lowest original gravity.

One must remember that Prädikat wine cannot be produced in every vintage. After a bad summer and insufficient autumn sun, the grapes may reach Kabinett but not Spätlese or Auslese quality. These are the years where fraud is committed, for it must be obvious to the authorities that the Prädikat wines presented for quality examination and an official examination number are not genuine Prädikat wines. Thousands of fake Prädikat wines have been passed as genuine by the examination boards.

A section of the Rheinhessen wine trade demanded that the basic condition for Prädikat wines, that they must not be enriched by the addition of sugar before fermentation, should be eradicated. I do not think that Germany is yet ready for this, but I do not exclude this possibility.

There are voices who wish to adopt the French system, i.e. one grower producing one type of wine. Here the grower gathers and crushes all his grapes at the same time, equalising the quality and producing just one wine. If you buy a Château Latour of a particular vintage you know that you buy the Château's only wine of that year. The German counterpart, Schloss Vollrads, does not only offer you Qualitäts and five different Prädikat wines, but also in some categories two different wines - the semi-dry and the medium sweet. If Germany would adopt this method, less doctoring of the wines would be necessary. The higher class Prädikat wines would improve the lesser ones, and the consumer would receive an equalised wine of middle-of-the-road quality, but which might still need sweetening with grape juice.

The wine trade and its journals engage in endless discussion on this subject. Headlines such as 'No Excuse for Sugaring' and 'Bitter Sweet Wine War' are commonplace. That sugar may be added in syrup form, that is dissolved in water, has caused particularly acrimonious discussion because the water dilutes the acidity and flavour, not to mention the other 400 constituents, even as trace elements. The practice should have finished years ago, but was extended to 1984 and there is no doubt that it will be extended again to 1990 by the EEC, though it is no longer allowed for all German regions, nor for all grapes. Wines from the less favoured regions such as the Ahr, Mittelrhein, Rheingau, Mosel-Saar-Ruwer and especially wine from the Riesling grapes (with its high acidity when not fully ripe) may be treated with up to 10% sugar solution.

'German wine is unique amongst all wines!' runs a German advertising slogan. It is indeed unique! German wine is enriched with sugar during fermentation, de-acidified, sweetened with grape juice and then ornamented with a beautiful label on which is printed the longest and most unpronouncable names of which only 75% need be true and in some cases is totally untrue. This is legalised fraud.

FRANCE

The addition of sugar in France has long been a contentious subject. According to P. Morton Shand, it was practised regularly at the Clos de Vougeot vineyards in Burgundy prior to 1790. The practice of adding sugar to a fermenting must is known in France as 'chaptalisation'. It is named after a famous French chemist, J. A. Chaptal, who lived between 1756 and 1832. He is sometimes referred to as 'the father of the sugar beet industry'. It follows that in France there is no objection to using ordinary beet sugar when permitted to do so and 25,000 tons of beet sugar are annually reserved for the wine trade.

The sugar is scattered over the red grapes during fermentation or dissolved in some of the juice and then added. The fermentation of 16.4–17.4 g of sugar in one litre of must will produce 1% of alcohol. It is permitted to add sufficient sugar to increase the alcohol content of a wine by up to 2.5%. It follows then that between one fifth and one quarter of the alcohol in a bottle of wine can be produced from added sugar.

Since the turn of the century, the practice has been steadily increasing in France, partly because the price of *vin ordinaire* is geared to its alcoholic strength and sugaring does not affect the flavour. In the 1920s, chaptalisation had become such a crying scandal in the Midi, where the grapes are so sweet that 13% of alcohol can be produced naturally, that the French Government passed an Act forbidding chaptalisation south of a line from Lyons to the northern limit of the Bordelais. To the north of that line, however, it is permitted to add as much sugar as will produce the 2.5% alcohol just mentioned. Generally speaking, however, the final alcohol content must not exceed the fixed amount.

IS IT REALLY NECESSARY?

Some wine lovers may deplore the practice of increasing the strength of a wine by artificial means, but it is not a scandal. It has to be admitted that in the northern areas of France as in all the wine producing areas of Germany, most of the wines produced exclusively by natural means, would otherwise be virtually unsaleable and undrinkable in seven years out of ten. Sugar is after all a natural substance of the grape. And France does not allow any sugar solution; the sugar must not be dissolved in water, but in the must or young fermenting wine.

In the last twenty years or so an alternative method to chaptalisation has been discovered, but it is, alas, expensive. The proposal is that the grapes in a poor year should be exposed to a temperature of 50°C (120°F) for a period of twenty four hours, during which time they would lose about 25% of their water, but nothing else. After pressing the grapes, the juice would contain sufficient sugar to produce an adequately strong wine by natural fermentation. The grower, it was quickly pointed out, would only produce three-quarters as much wine and at an additional expense in terms of space, labour and energy. It is simply cheaper and less troublesome to add beet sugar!

ALCOHOL, TOO

Although it is not common practice to do so, a certain 'enrichissement' of wines occasionally occurs. This is the addition of some alcohol to an ordinary wine of low alcohol content to make it stable. This is not the same as 'fortification', and the conditions - kind of alcohol, category of wine, quantity etc. are laid down in EEC regulations.

In Spain, grape brandy is added to sherry after it has been fermented, to increase its alcohol content to around 17-20%. This is primarily for the British market, where, through ignorance, it is often left opened in a decanter for weeks or even months or left in a half empty bottle. Without the fortification with brandy the wine would quickly oxidise and spoil. In Spain they drink their sherry and montilla wines unfortified. In Portugal, too, grape brandy is added to port wine to increase the alcohol content to around 20% and to inhibit fermentation, so leaving a sweetness of residual grape sugar. Both countries specify the use of grape brandy and forbid the use of wood alcohol or rectified neutral spirit which is cheaper. Spirit is also added to the poor quality wines that have been flavoured with herbs and spices to make Vermouth.

11

The Liquid Sugar Scandal in Germany

As I have already mentioned, Germany's vineyards are among the most northerly in the world. Although the publicity people use phrases like 'pampered by the sun' to promote the wines produced, the fact remains that in at least seven years out of ten there just isn't enough sunshine to ripen the grapes fully. As a consequence it is essential to add some substance to balance the acidity and to increase the alcohol content so as to make the wines palatable.

The EEC and German regulations have fixed minimum alcohol contents for Tafelwein and Qualitätswein, very low potential alcohol for both, and also minimum actual alcohol for the wine after enrichment with dried sugar.

After the enrichment, and when fermentation is complete, all German wines are bone dry, with the exception of a few Auslese, the Beeren and Trockenbeerenauslesen. To make them drinkable, all wines including those which will be marketed as Trocken (dry) receive a certain amount of süssreserve (unfermented grape juice) just before bottling. This is not so easy and is a quite expensive task for the grower. He has to store and prevent the unfermented grape juice from fermenting until bottling time. Grape juice also has a great disadvantage; a relatively large quantity is needed to increase the sweetness of the wine, but the increased volume reduces the alcohol content which may fall below the permitted minimum. Liquid sugar is a more concentrated form of sugar and the final product presents itself even better although it is cheaper! A very welcome alternative for the grower – but illegal, because it adds water to the wine and so is forbidden by law. Alas, temptation abounds, and in the dim light of the cellar it is difficult to see that the additive is very often not süssreserve but liquid sugar, similar in appearance, extremely difficult to detect, much cheaper to use, and capable of doubling the price that the wine would otherwise fetch. German growers and co-operatives did not care – even some of the 'greatest' fell to the temptation. The result was 2,500 criminal cases in which there were often two, three or even more people involved.

For the Germans, an offence against the wine laws was considered a *kavaliersdelikt*, a cavalier or nobleman's game. Like a duel, it was just a crime that did not spoil an honest name! But in the liquid sugar cases, as far as they were dealt with in open court, that was not a *kavaliersdelikt*, – it was fraud.

In the autumn of 1980, major investigation uncovered a widespread practice of enhancing the quality of the wine with the aid of liquid sugar – a solution of fructose and glucose known as invert sugar. One ton of invert sugar is sufficient to improve the quality rating of 7,070 gallons of Tafelwein to Auslese. For those who prefer simpler figures, this is equal to 220 g (7¾ oz) in each 70 cl bottle!

In the Rhineland Palatinate, Herr Werner Hempler, the chief prosecutor of the central authority for offences against the wine and food laws, stated on 31 October 1980 that 200 liquid sugar merchants and some 1800 wine growers had been found guilty of the illegal use of sugar. In the Palatinate some 6,270 tons of liquid sugar had been used in the years 1977 to 1979 – enough to sweeten some 300 million bottles of wine. Herr Hempler went on to state that since 1978 merchants had sold 4,070 tons of invert sugar in the Mosel-Saar-Ruwer area; 550 tons in the Nahe; 1,150 tons in the Rheinhessen; 510 tons in the Rheinpfalz; 3.5 tons in the Mittelrhein and 4.4 tons in the Rheingau. Substantial, although unquantified amounts, were thought also to have been sold in Luxembourg and Alsace. To quote a case from open court: a co-operative in the Palatinate found when gathering the grapes that many bunches had fallen off the vine because of bad weather. A quantity producing 40,000 litres of must had to be lifted from the ground. These grapes were far from ripe and the wines, even after the permitted enrichment with dry sugar, were thin and poor. The permitted addition of grape juice would reduce the alcohol content too much, but liquid sugar would give the same sweetness and leave more alcohol in the wine.

To buy the necessary liquid sugar and add it to the wine involved two transactions which if entered as such in the co-operative's ledger and wine cellar book would be sufficient documentary evidence to send the cellarmaster and his manager to prison. The work was performed by the cellarmaster entirely on his own and in the absence of all other personnel, in the evening and during the night. The books and vouchers were falsified, and even the suppliers had, according to their books, never sold a drop of liquid sugar. In their books the liquid sugar was entered as 'chemicals to fight pests in the vineyard'!

This more or less gives the general picture of the transaction and use of liquid sugar. In many cases liquid sugar was used to bring the wine up to a higher class, a Kabinett wine becoming a Spätlese, and a Spätlese an Auslese, and so on.

The penalty for offering adulterated wine for sale is up to three years' imprisonment and the liquid sugar merchant can be fined up to DM 50,000. The risk of detection is not great, however, and the proceedings, when the culprit is caught, are more or less in secret and may end with only a negotiated fine, depending upon the seriousness of the offence.

The last case heard by the Mainz Court was that of the former President of the German Winegrowers' Association, General Werner Tyrell, who was able to continue this illegal practice throughout the three years of investigation and prosecution of a total of some 2,500 wine growers, including himself, until 1983. I had actually drawn the attention of the Committee of the Rheinland-Pfalz Parliament investigating the liquid sugar activity to the 'great ones' and had even mentioned his name to them.

If you read the speeches of this man during his many years as leader of the wine growers, it is astonishing to see how the 'great' and 'richest' can act freely and illegally without risk of investigation. Why the leader of the Association was the last to be prosecuted is not known. General Tyrell was a prominent member of the association of producers of Prädikatsweine and was praised for his authentic and unadulterated wines! Compared with some of the others who were prosecuted, the President got off lightly. He had confessed at the last minute and this was used in his defence. He was sentenced to imprisonment for one year but, of course, it was a suspended sentence! And the greatest part of the profit remains in his bank account.

The most astonishing part of the hearing, indeed of any hearing in my memory, was the question that the Public Prosecutor put to General Tyrell. In Germany the prisoner at the bar is not heard under oath. The rationale of the law is that one cannot expect a prisoner to speak the truth and it is better to allow him to lie than have daily perjuries. The question asked by the Public Prosecutor was: 'Will Herr Tyrell make this statement under his word of honour as a German officer?' That is the mentality of people who have to administer the law. And this question was asked after Tyrell had admitted his 'responsibility' and after his cellarmaster had stated that what had been done in Tyrell's cellars happened acording to the orders given to him.

This case brought to light another important revelation. The 'Chemiedirektorin' of the Wine Control, 47-year-old Brigitte Holbach, stated that she had informed the Head of the Wine Control as early as 1979 that she had found sweetened Prädikat wines among the Tyrell samples examined, but she was prevented by her superiors from informing the Public Prosecutor of her findings. After she had given her evidence she was threatened in an anonymous midnight telephone call that a newly formed 'Club' would murder her!

From time to time Germany tries to set an example just to show the growers and traders that not everybody can get away with a fine negotiated in secret. Such cases are restricted, however, so as not to do damage to the good name of the trade and not to do damage to the German economy. This is German justice! A case in which the judgment was particularly severe was reported in the *Off Licence News* on 1 August 1985. The Schmitt brothers, who owned one of the largest wineries in the Mosel-Saar-Ruwer region, were held in jail for approximately eighteen months before their case was heard. At the hearing the Prosecutor estimated that the winery which produces around 2.5 million cases of wine a year illegally manipulated more than 1.1 million cases of wine between 1972 and 1975.

The firm's crime included using 231,000 lbs of liquid sugar, labelling and selling chaptalised wine as German Qualitätswein mit Prädikat and fraudulently accounting for it as vineyard and cellar supplies.

The Prosecutor also estimated that the company made about £3 million in illegal profits by doctoring low quality grapes to bring the wine up to the standard of table wine, and tampering with table wine to upgrade it to quality wine and manufactured Prädikat levels. The brothers were fined DM 20,000 (around £5,000); Heinz-Gunter Schmitt was sentenced to five years' imprisonment and his brother Gerhard to four years. The managing director received a two year sentence and the cellarmaster a year and nine months. Whether other growers have heeded the warning is open to doubt.

12

Liebfraumilch is Lovely

Liebfraumilch is a name that stands out on all wine lists outside Germany, a name that many foreigners identify with German wine. It is often the only German wine they know and drink. Contrary to popular belief, Liebfraumilch is not a district at all, but an invented name which to 1971 could be applied to any pleasant wine of good quality. The name is derived from the Liebfrauenkirche (Church of Our Lady) at Worms, which is surrounded by vineyards. Liebfraumilch was originally Liebfrauminch. 'Minch' is an old word for 'moench' meaning monk. Liebfrauminch wines, therefore, were made by the monks at the Liebfrauenkirche. The natural development of the language has changed the consonant 'n' to 'l' and 'Liebfrauminch' to 'Liebfraumilch'. Liebfraumilch was always a blended wine. Before 1930 it often contained up to 49% of French or Spanish sweet wine, to give it roundness and a touch of sweetness, more alcohol and staying power. When blending of German white wine was prohibited in 1930, the blend consisted of Rheinhessen, Rheinpfalz, Mittelrhein, Nahe and Rheingau wine.

From 1871 to 1918, and again from 1940 to 1945 when Alsace was incorporated into Germany, much wine from this country was blended into Liebfraumilch. From 1937 to 1945, when Austria was the German Ostmark, Austrian wines were also used to blend with Liebfraumilch.

Fancy names can be used for wines of any origin and the name or some similar name such as 'Liebchenwein' was used in the USA and Australia until stopped by litigation. It is still used and cannot be stopped because of long usage in some other countries.

Since the Wine Law of 1971 the 'legal' content of a bottle of Liebfraumilch has undergone various changes. I do not intend to go into all the details, but to emphasise the scandal that happened in the development. The 1971 Law had established that only 'type wines' of regional origin could be created. The Federal State of Rheinland-Pfalz imposed a number of conditions for the use of the name Liebfraumilch and in particular that:

1. It had to be Qualitätswein.
2. It had to originate in Rheinhessen, Rheinpfalz, Nahe or Rheingau.

This was interpreted as meaning that Liebfraumilch should be a blend of wines from the four regions, an interpretation which is not in accord with the EEC regulation that a quality wine must originate in only one region.

Rheinland-Pfalz, the German Federal State that bothers least - whether right or wrong - as long as the voters don't revolt, did not just condone, but positively encouraged this illegal action. As one lawyer remarked, 'There exists in Mainz an unwritten law - do as you like!' Mainz is not only the seat of the Government of Rheinland Pfalz, but also the home of the famous Mainz cheese. The lawyer's view is often more dramatically expressed in the phrase, 'In Mainz stinkt nicht nur der Kase', the sense of which is perhaps best expressed in the English saying, 'There is something rotten in the State of Denmark.' The leading shippers pleaded for the creation of a new region for Liebfraumilch, in addition to the eleven regions already existing, which would include the four specified Rhine regions given above. The only wine of this twelfth region would be Liebfraumilch; the designation of the four regions amalgamated in it would continue separately as before. It was a most unacceptable solution and was rejected.

It has now been confirmed that Liebfraumilch can originate in any of the four regions and that will be four types of wine, Liebfraumilch Rheinhessen, Liebfraumilch Rheinpfalz, Liebfraumilch Nahe and Liebfraumilch Rheingau. The shippers objected to this ruling, claiming that it would be impossible to fulfil their obligations.

The quality of wines produced by different shippers and marketed as Liebfraumilch can vary enormously, not only between shippers, but also between bottles from the same shipper, and bottles bought at different times. Liebfraumilch is a blend of different quality wines produced in one of the four regions mentioned and what one blender regards as a harmonious wine, another might totally reject. The constituent wines in the blend are constantly changing too, not only from harvest to harvest but also by marketing success. If all the stock of one constituent is used, it must be replaced by another and no two wines are ever the same. Liebfraumilch is never a great wine although some advertisers would have you think it so. I pity those people who all their lives drink only the same Liebfraumilch. I say to them, 'Buy two or three or even four different brands and taste them. Choose the one you like best and then taste this against another few bottles of different brands. This expense may save you a lot of money in the years to come and you will be drinking the wine you like best.' The late and much respected Otto Loeb refused to import and sell Liebfraumilch. When some customers insisted on Liebfraumilch he supplied them with a wine of geographical origin labelled Liebfraumilch, Oppenheimer Goldberg.

I have had grave doubts when tasting some Liebfraumilch wines. On one occasion I had lunch in the office of a well known shipper and he gave me his Liebfraumilch brand. I told him that without the label and if I had been asked to guess the origin of the wine, I would first of all have given the name of the grape – Gewürztraminer. The bouquet was so overwhelming that it was very difficult to guess at any geographical location. The occasion made me curious and I subsequently bought quite a few bottles of his wine with different examination numbers, but I never came across the same bouquet! He did not have the courage to give me the actual wine of his famous brand, or perhaps he just wanted to examine my expertise. Was he afraid of my outspoken criticism? On the other hand, this small fiddle confirmed my general view about the brand.

The *Decanter* magazine (which gives great service to wine lovers) arranged a tasting of a dozen different Liebfraumilch wines. The panel for a *Decanter* tasting always consists of wine merchants, connoisseurs and amateurs, some wine writers and as a rule includes a Master of Wine. The most expensive brand of Liebfraumilch and the one with the greatest turnover got very low marks!

Why drink a German wine with a fancy designation when you can get a bottle of wine with exact geographical designation often at a lower price? Under the name of Liebfraumilch many sins are hidden. A great part of Germanised wines are exported under the Liebfraumilch label. The exporters who create a large demand for their brand name, export millions of cases of wine each year and cannot ensure, by tasting and comparison of the different constituents, that their brand quality remains the same. They just accept what wines come along. The best example of this was from a former Minister of Viticulture who acted as a middle man. He made an offer, to a leading Liebfraumilch exporter, of wine that he had received from a broker who had in turn received it from another broker. The offer was accepted and the wine was delivered in large wine tankers; 200,000 litres of the wine were lying in the exporter's cellars ready for bottling and distribution under his famous label when the wine was blocked and later confiscated. It turned out that the wine had come in the same tankers from Italy and was merely Germanised by the blending in of a small quantity of German süssreserve!

In conclusion, Liebfraumilch today is not what it was in 1929, 1931 or 1969. Today there are four types of Liebfraumilch from the four specified regions. There is a mandate that 51% of the wine must originate from Riesling, Silvaner, or Müller-Thurgau grapes, and the end product must have the character of the grape. As the demand grows, the use of other grapes will shortly be allowed. In any case Riesling wine of any origin is in such demand that better use can be found for it than blending it in Liebfraumilch. Since more than 50% of German wine exports bear the name Liebfraumilch, it is hardly to be expected that a person drinking identical label Liebfraumilch gets the identical wine, especially if the shipper endeavours to sell tens of millions of bottles annually. Liebfraumilch is never a great wine; at its best it could be a thirst quencher, but it must be 'Lieblich'; that means that it contains 24 g of unfermented sugar!

Liebfraumilch is an export wine; a German native will not touch it, and it is certainly not available in German hotels other than the few where English tourists ask for it. It is, after all, a name so easy to remember and to pronounce!

13

A Horse and Nightingale Pie

THE 'BERNKASTELER DOKTOR' STORY

The name of the site 'Bernkasteler Doktor' is attributed to a legend. Archbishop Boemund II, Elector of Trier 1351–1362 (so runs the tale), lord of large tracts of vinelands on the rivers Mosel and Rhine, and owner of a castle situated on the windy heights above Bernkastel, was addicted to spending much of his time in this favoured spot. One day he fell ill there. He was grievously sick and his doctors plied him with medicines in vain. As he lay on what was thought to be his deathbed, a flask of wine was brought to him from one of the best sites in the neighbourhood. He drank, and a miracle was wrought. The dying prelate recovered and in gratitude bestowed on the wine - and the site - the appellation 'Bernkasteler Doktor'.

'Ye who are sick and sorrowful,
Arouse yourselves and take a pull
Of Wine - the finest "Doctor".
It's better than the best of pills,
For "Doctor" Wine can cure all ills -
A great and kindly doctor!
For cheering draughts so justly famed,
Its native hill is proudly named,
Just like the Wine, "The Doktor".'

The size of the Doktor vineyard was actually only 1.35 hectares (3¼ acres). There is so little real Doktor wine produced and the demand is so great that two of the three growers to whom the Doktor vineyard belongs, gather the grapes with adjoining vineyards and market it as Bernkasteler Doktor und Graben (Thanisch) and Bernkasteler Doktor und Bratenhöfchen (Lauerburg). The third grower has the same procedure for a part of his crops:

Bernkasteler Doktor und Badstube (Deinhard). Thanisch owns 40.6% of the original hill, Lauerburg 4.0% and Deinhard owns 55.4%.

No German wine is so over-rated as the Doktor. In blindfold tastings, wines from adjoining vineyards of equal quality cost only half the price. The cause of all this is the demand for Bernkasteler Doktor from the USA, and one wonders just how much wine from the real Doktor vineyards each cask or bottle contains.

One is reminded of the tale of the restaurateur who sold a nightingale and horse pie - and when interviewed by a food inspector regarding the proportion of each component, answered 'half and half - one nightingale and one horse.'

A site may only be registered in the German Vineyard Register if it is at least 5 hectares in total area. Smaller areas may only be admitted if the formation of a larger site is impossible, either due to the use of the neighbouring land, or because of the very special character of the wine produced from the vineyard in question. The Bernkasteler Doktor is one of the most renowned individual sites in Germany and it was therefore proposed that the site should be entered at its original size of 1.35 hectares.

The 1971 Law, however, prohibited the naming of a wine as originating in two sites - as was the case here with Doktor and Graben. If the wine contained 3 measures of Graben and one of Doktor, then legally the name of the wine should be described as Bernkasteler Graben. All this did not disturb the Thanisch Estate, and the Public Prosecutor's Office in Trier accepted the view that Doktor and Graben did not represent the name of a blend but the name of one vineyard. This, in any case, was the gist of the owners' application - to enlarge the Doktor site just to the size of their own property. From about 5,000 square metres Doktor and 14,500 square metres Graben, a total of 18 Fuder 'Doktor und Graben' were being marketed. It is easy to see that the contents of the bottle are in opposite proportion to that implied by the wording on the label.

The growers of Bernkastel and the other owners of vineyards on the Doktor slopes took up this challenge. Deinhard wanted either the Doktor site to remain unchanged, small though it was, as an exceptional case; but if an enlargement was favoured then their Badstube site should be amalgamated in the Doktor so that in the final result they would still own 55.4% of the enlarged Doktor site.

Neighbouring growers, on both sides of the Mosel, now made

known their demands - based on the quality of their wines. Their vineyards produced wines quite as good as the Doktor blends sold in the past and, they therefore contended, their vineyards should be included in any new arrangement.

The Ministry of Viticulture, who had the final decision, agreed with the latter opinion; the Doktor site was extended to the new size of 5 hectares (12 acres), which increased the number of owners from three to thirteen, including a charitable institution, who had leased their vineyard to various growers. Thanisch have appealed against the Minister's decision with success.

The Court had allowed the new owners to label their wines 'Doktor' while the case is pending and new owners have offered their wines - as far as available - from 1970 onwards as Doktor. Negotiations have been going on in an endeavour to come to a friendly settlement, but whenever it comes to putting signatures to an agreement everything stalls. The final judgement of the administrative court in Koblenz accepted the Thanisch arguments and fixed the Doktor vineyard in Bernkastel at 3 hectares.

The potential price of a lease on the Doktor has increased many times. New owners offered their Doktor at three times the former price of wines under different names from the same spot. For example:

	DM/doz		DM/doz
1977 Doktor Kabinett	190	- Wehlen Sonnenuhr	67
1978 ,, ,,	200	- ,, ,,	72
1979 ,, Auslese	324	- ,, ,,	135

The new Doktor may consist of 75% Doktor vintage, plus 10% unfermented grape juice and 15% Qualitätswein from Mosel-Saar-Ruwer (a blend from the vineyards between Luxembourg and the Rhine).

Or to put it poetically 'A Horse and Nightingale Pie', at least for people who buy their wines only by the names on the labels - snobs but not connoisseurs - and who is to blame? Konrad Adenauer, of course, who brought General Eisenhower a case of an extraordinarily fine Doktor as a present and thereby stimulated a demand for wine with the same label from people willing to spend so much money believing that they buy the same quality! I would rather spend the same amount of money on three bottles from a reputable grower and have the same pleasure three times.

14

The Wine Laws of Europe

EEC LEGISLATION

Fundamentally these laws consist of regulations on the common organisation of the market, establishing certain standards of quality that are readily recognisable. Each wine producing country has more precise regulations controlling the variety of vines that may be planted, the number of vines that may be planted per hectare (2.47 acres), the quantity of wine that may be produced per hectare, the amount of sugar, if any, that may be added during fermentation, and so indirectly controlling the amount of alcohol that may be formed, the oenological practices which are allowed, the nature and quantity of the chemicals which may be added (e.g. sulphur, finings, etc) the quality to be attained to recognition and so on. Each country is naturally anxious to raise the quality of the wines it produces and its internal laws are designed for this purpose, but the EEC laws are stronger than the national laws and the latter must not contravene the EEC laws. It is when these laws are broken that scandals occur.

The Council of the EEC Regulation 355/79 laid down general rules for the description and presentation of wines and grape musts, making it quite clear that potential buyers and public bodies responsible for organising and supervising the marketing of the products should have information which is sufficiently clear and accurate to enable them to form an opinion of the product. And the rules should ensure that this purpose is served. Therefore some mandatory information must be given to identify the product and optional information to indicate the special properties of the product or to characterise it. In addition to that, the EEC Commission Report No. 997/81 implementing the Council's regulations, gives detailed instructions about labelling, and in some cases even the size of print to be used.

Table wines

The EEC laws distinguish three different kinds of table wines:

(1) The table wine of a member state made from its own grapes.
(2) Table wine produced in one member state but made from grapes grown in another member state.
(3) Table wine resulting from a blend of different wines originating in more than one member state. The words 'wine from different countries of the European Community' must appear on the label.

It is the latter category which is in the centre of the great scandal, the so called 'Euroblend', a (term first used by my son, Peter). There are still some growers alive who remember the German wines from before 1930, when the sterilisation filter was quite new and German wines were made sweeter and fuller and the alcohol content was increased to give them more staying power, by blending them with 49% wines from France, notably Sauternes and Montbazillac, or else with sweet wines from Spain. When the EEC laws again allowed the international blend, the Euroblend appeared pretty soon on the market, nicely dressed in German uniform, the Rhine or Mosel bottle, labels often showing well known German landscapes or even villages and a good German brand name, everything of course printed in gothic letters. To the viewer, the perfect teutonic knight. And in shops they were standing amongst German bottles and in price lists under the German head line. The ordinary buyer and salesman did not even know the difference.

Table wine must not bear any geographical name even in the address of bottlers. They are replaced by a code to assist any misunderstanding in the mind of the buyer, a regulation which is often not considered, the full name and address of the German bottler and the additional 'packed in Germany' or 'bottled in Germany' makes the buyer believe that he drinks a German wine, which in some cases appears in special offers even as 'Liebfraumilch type'.

This scandalous merchandising has led to a tremendous increase in turnover, not of the Germanised wines, but of wines passed off as German wines. Nearly 25% of German exports are not German wines but wines exported from Germany. The USA has taken the necessary steps to protect American wine drinkers: the origin of the wines in the Euroblend had to be stated in percentages, enumerating them downwards; for example, Italian 95%, German 5%.

Unfortunately the German government does not regard the German exporters as trading unfairly and expects the countries of distribution to act. Europe may adopt the USA rules. It is up to the wine drinker to use and read the information which is given and not to buy blindly. These wines are often quite good and good value for money, but certainly not the wines for the person who wants to drink a German wine.

Quality Wines

The labelling of quality wines must include the following information:

1. The specified region of origin.
2. The traditional form of indicating quality wine in the various member states; e.g. in Germany, Qualitätswein or Qualitätswein mit Prädikat; in France, Appellation d'Origine Contrôleé (AOC), or Appellation Contrôlée Champagne, or Vin Délimité de Qualité Superieur.
3. The volume of wine in the bottle.
4. The name and business name of the bottler and the local administrative area or part thereof in which the head office is situated.

The law permits additional information for table and quality wines, especially the colour - red, white or rosé; a brandname; the vintage; and other information. The use of brand names or rather the German attitude regarding brand names has led to litigation in Germany, Great Britain and the USA. It is laid down that products, the description or presentation of which does not conform to the regulations may not be held for sale, put into circulation in the community or exported. The member states have appointed bodies responsible for ensuring compliance with the provisions of this regulation.

UNITED KINGDOM

Until its admission to the EEC, the only national laws relating to wines were contained in the Merchandise Marks Act of 1887 and the Trades Descriptions Acts 1968/72. The making and selling of 'wine' from raisins had been permitted since 1635 and from concentrated grape juice for approximately the last one hundred years. This was well known to the wine trade and accepted by them as harmless competition. Customs and Excise Officers derogatorily describe these wines as 'sweets' but 'British wine' has been accepted as 'made wine' and rightly so. After all, it is based on grapes, albeit not freshly gathered, but the concentrated grape juice is restored to a near normal gravity before fermentation. The end products are clean, alcoholic beverages which satisfy those who buy them. Although wine making of this kind is prohibited by the wine producing countries, these wines are a very good means of introducing wine to those people who cannot yet afford or appreciate a better wine.

Upon joining the European Economic Community, the United Kingdom had to recognise the wine laws of member countries, but a period of grace was given that expired on 31 December 1976. The United Kingdom wine trade as a whole has a pretty bad record in spite of the efforts of the various Trade Associations. According to the general opinion, Nuits-St George was a generic name for a Burgundy with a good body. The 'Ipswich Affair' (see page 154) was still the exception. From 1 January 1977 the United Kingdom wine importers and merchants, as well as the increasing number of English wine growers, have had to conform to the EEC laws controlling wine.

FRANCE

France was the first country to enact comprehensive regulations about its wine as far back as 1934, although individual laws have been effective for many years. A system called Appellation d'Origine Contrôlée (AOC) was devised just 50 years ago. This provided for the issue of a certificate of authenticity to a wine that complied with all its local regulations. It entitled the producer to advertise, usually by printing the name of the local vineyard, village or district in which the vines were grown and the wines were produced, followed by the words 'Appellation Contrôlée'. In the course of time some 400 different named vineyards, villages or districts have been defined, and an additional 80 for wines of superior quality (*VDQS*) and *Vins de Pays*.

Until the summer of 1975 a vineyard, village or district producing more wine than that authorised in the regulations concerning that area could sell the identical surplus without a certificate of authenticity, that means as table wine and without any statement of its geographical origin. This was more a marketing than a quality regulation which many countries did not recognise. The United Kingdom was one and bought many such wines. Because they were of the same quality as that which had a certificate, UK wine merchants marketed the wine with the designation of geographic origin and should not have added AOC, but not all merchants kept to this.

Generally speaking, the more wine produced from a given number of vines, the lower the quality. Production limits are set for each designated area in an effort to raise and maintain the quality of the wine. Deliberately exceeding these limits was and is an offence, and the wine authorities in France have long been concerned with this problem. From the harvest of 1975 the regulations were tightened. A producer could keep within his quota and sell all his wine with a certificate of authenticity, or produce as much as he wished and sell it without a certificate at all, or sell his quota with a certificate and any excess for distillation at one franc per litre.

As with all regulations, some people feel strongly opposed to this arrangement. They argue that since the quotas were fixed, the methods of production have improved considerably and the quotas should now be increased anyway.

Wines that do not reach the standards prescribed for an AOC certificate may reach a somewhat lower standard such as Vin Délimité de Qualité Supérieure (VDQS). This is usually a blended regional wine and is generally speaking very good value for money. Obviously less expensive than an AOC wine, it may come from an area not covered by AOC regulations, or from vines not authorised for the AOC area from which it comes.

Lower still in the scale is Vin de Pays. This was usually a blended regional wine with the general characteristics of the region from which it comes. It was usually thought to be a little better than Vin de Table, more technically described as 'Vin de consummation courante'. This was no more than a blend of wines from anywhere in the EEC, mixed with poor quality wines from France, and as its name suggests, it is for current drinking and not for keeping. It is often found in French cafés and served by the carafe. 'Vin de Pays' is today a better regional wine that is also exported.

ITALY

In 1963, Italy established laws on similar principles to the French, but more precisely in line with the Common Market regulations that had been agreed in 1962. The principle of associating a wine with the area where its grapes are grown was maintained and the regulations are known as Denominazione Di Origine Controllata or DOC for short.

The regulations are administered by the National Committee including growers, wine producers, brokers, technical experts and members of the National Union of Consumers. They employ officials who are constantly checking to ensure that the regulations controlling the growing, production, maturing, bottling and labelling are fully observed. As a result one can be sure that a bottle of wine bearing a DOC label:

1. comes from the area named;
2. is produced from the prescribed proportions of specific grapes, by the traditional methods and has been properly matured;
3. that the vineyards have been inspected and that the maximum grape production and wine yield has not been exceeded;
4. that the sales of the wine have been recorded in ledgers subject to inspection;
5. that the vintage year on the label is accurate and that the label is not misleading in any way;
6. that the wine has met the Committee's high standards of taste, flavour, bouquet, colour, chemical composition, alcohol and acid content.

There are three levels within the DOC regulations:

1. *Denominazione Di Origine Semplice*: a blended wine from a particular locality.
2. *Denominazione Di Origine Controllata:* similar quality to the French AOC wines. Some 200 certificates of authenticity have so far been granted.
3. *Denominazione Di Origine Controllata E Garantita:* wine of superior quality that has been bottled in the area of origin in a container no larger than 5 litres. By 1977 no wines had been so classified, although Barolo, Barbaresco and Chianti had long been mentioned as probable candidates and have now been classified together with Vino Nobile Di Montepulciano and Brunello Di Montalcino.

Italy is the largest wine producing country in the world. Some 1,500 million gallons of wine are produced of which only 200 million gallons are exported, mostly to Germany and France. Much of the wine exported to the United Kingdom is Italian bottled and of DOC quality.

THE NEW WORLD

Although 75% of the world's wine is made in Europe, Australia, South Africa, South America and USA are fast becoming substantial producers and exporters of wine. So far only South Africa has developed a system of named origins and registered estates to indicate the authenticity of their best wines, which it has operated since 1972.

The Australian wine experts frequently talk about the subject of Appelation Contrôlée, but their country is so vast and the range of wines made at each winery is so diverse that no one has yet come up with a workable system. The winemaker often prints many details of a premium wine on its label. This is likely to include the name of the grape(s) from which it was made and where they were grown; the year of the vintage; the maturation period in oak; the alcohol content and a somewhat poetic description of the wine.

Three small and relatively new areas (Margaret River in Western Australia, Mudgee in New South Wales and Tasmania) have introduced a full scale system, complete with tasting panels, numbered certificates, stock checks, and the strictest requirements for district propagation of grapes and wine making. It is possible that these regulations will spread to other isolated areas but it seems unlikely that the system will become universal, since grapes are frequently grown in one area and trucked several thousand kilometres to a winery where the grapes will be crushed and turned into wine.

Most USA growers of repute name the origin of their wines on their labels, e.g. Napa Valley or New York State. After long debate regulations were introduced in August 1978, some of which came into force only in 1983, and these are being currently developed still further.

At the height of the 'anti-freeze' scandal in the summer of 1985, the Austrian authorities promised that the strongest wine laws in the world would be introduced. As if the wine laws could change the minds and mentality of people!

The country with the strongest wine law of all is Israel:

1. Wine has been of great importance to Jews since the time of Moses. For religious reasons, therefore, it must be clean (Kosher) and nothing may be done to detract wine from its natural development and its true nature.
2. The number of firms or co-operatives allowed to produce and treat wine is restricted to eight.
3. Supervisors have been appointed for each cellar of the eight wine makers and the entrances to the cellars are secured by two locks - one for the owner and one for the supervisor - therefore there can be no activities in the cellar in the absence of the supervisor!
4. These are in addition to the Kosher wineries and monasteries producing and bottling wine - they are under their own ecclesiastical supervision.

During the Middle Ages in Germany, the Kosher wines were prepared by the wine lovers and considered the best. It pays sometimes to try the unknown. The Israel vintage Cabernet Sauvignon - the best of all Israeli wines - is well worth trying, so is the Grenache Rosé - my two favourites. The varietals are improving every year, as the young vineyards with a little more age produce firmer wines.

There are now in France and Italy a few firms producing Kosher wines under the same strict conditions and supervision as the cellars in Israel. There would have been no question of adding diethylene glycol, as it is forbidden to use many of the EEC permitted substances such as casein or blood.

15

Oenological Additives and Practices

Wines of former years were not often as good as the wines made today. Two centuries ago, imported wines were often blended with English home-produced fruit wines to refresh them and make them saleable. Many imports came from Bordeaux or Cadiz in small sailing ships that rolled and tossed in the rough waters of the English Channel. Others came by barge down the Rhine to Antwerp and then by ship across the North Sea and up the Thames to London. It was hard on the wine and many were served both sour and cloudy. Gradually improvements have been made and wines today are much better handled. But they are no longer the naturally fermented juice of freshly crushed grapes.

The vines now receive frequent sprays with sulphur and other chemicals to prevent the growth of fungi that damage the plant and diminish the crop. When the grapes are crushed they are mixed with liquid sulphur dioxide or a sulphite powder to kill off the bacteria, wild yeasts and moulds that form the bloom on the skin. The juice may then be chilled to encourage solid particles to settle, from which the clear juice is then drawn off. Or it may be centrifuged to achieve a similar result.

Sometimes it may be necessary to reduce the acidity of the juice with potassium or calcium carbonate (chalk) or to increase it by the addition of tartaric acid. Red wines may also need the addition of some tannin to provide firmness and character.

In many wineries sugar will be added so that sufficient alcohol can be produced to make an acceptable wine. The quantity varies with the local laws and the lack of sunshine in a particular year. It can vary between enough to increase the alcohol content of the wine from by 2% to 4.5%. A pure yeast culture may then be added together with ammonium salts to provide essential nitrogen for the yeast. Vitamin B1 in the form of Thiamine may also be added to nourish and stimulate the yeast.

Fermentation may be regulated by cooling the must and therefore slowing the action to enhance the flavour of the finished wine. Next the young wine must be removed from the sediment of dead yeast cells and other debris. White wines may be passed through charcoal to remove colouring matter. Nearly all wines have some fining agent added to them. This may be isinglass made from the air bladders of sturgeon, animal albumen made from eggs or dried blood, casein, a constituent of milk, various earths such as bentonite and kaolin, or a pectinolytic enzyme to break down the pectin that causes haze.

It is almost certain that sulphite salts releasing sulphur dioxide will again be added to prevent oxidation and infection of the wine. The wine may also be filtered to remove every cell of yeast or bacteria and so prevent further problems. Alternatively, potassium sorbate may be added to kill the yeast. In some countries - notably Germany - unfermented grape juice may be added to sweeten the wine. The wine may then be chilled to just above freezing point to precipitate any excess of tartaric acid in the form of tiny glass like crystals that might otherwise detract from the wine's appeal.

At bottling time the wine may be heated to 60°C/140°F for a few minutes to ensure stability. If the wine comes into contact with any metal other than stainless steel, the acids will combine with the metal to form a salt that imparts an unpleasant taste and cloudiness. Treatment includes the use of ferric-ammonium-sulphate and potassium-ferro-cyanide. Treatment for metalic 'casse' is essentially work for a qualified chemist, for should it not be carried out perfectly the wine would contain a deadly poison.

In 1935 very shortly after setting up my own bottling plant, I discussed the treatment of German wines with my friend, Victor Turner. He had a problem on his hands. He had supplied a Bordeaux wine to a good friend and customer and the wine was so cloudy that the customer threatened to return the wine. The contract for the wine and the old friendship were at risk. My advice was sought and I sent samples to the Viticultural College in Geisenheim. I soon received a report that the wine contained an excess of copper and iron salts and needed treatment with 'blue' finings - a name given because the fining makes the wine look blue before the metal is precipitated and falls to the bottom of the cask. The quantity of potassium-ferro-cyanide to use was given and I carefully wrote down the technical details for my friend.

Even so, he urged me to go with him to his cellar and supervise the fining by his cellarman.

On my arrival I found everything prepared but when I examined the fining agent I noticed that ferric-cyanide had been supplied instead of ferro-cyanide. The local chemist took some convincing that the substance he had supplied was wrong and must be changed. My work with the correct material was successful, the wine remained bright and of excellent quality. But if Victor Turner had not insisted upon my presence he might have poisoned a whole town with ferric-cyanide!

This is not the place to relate in detail all the ingredients and procedures that are legally permitted and prescribed by law, but a copy of the EEC Regulation 337/79 is given as an Appendix since it may be of interest to a few people.

16

What's in a Name?

SCANDALS OF FALSE LABELLING

Probably the greatest single cause of scandal has been the labelling of the bottles with intent to defraud or mislead. In the UK the Merchandise Marks Act and the Trades Description Acts apply and the Department of Trade and Industry are continually trying to find more precise ways of labelling wines. Their work and that of other European countries is being co-ordinated by the Wine Committee of the EEC. It is becoming increasingly difficult to find loopholes in all these restrictions, but nevertheless there was in 1976 a classic example of labelling with intent to defraud the buyer.

The Krug Case

The House of Krug has long been renowned for the high quality of its champagne. It is made by well established, indeed old fashioned, methods by today's standards. The firm refuses to use glass or stainless steel containers and prefers wooden casks both for fermentation and for maturation. They prefer to buy the best quality grapes available year by year rather than to contract to buy all their grapes from the same source regardless of seasonal variations. With high quality goes a high price and as a result they sell a large part of their output by direct mail. Krug reckon that they know most of their customers personally. Apart from certain high class restaurants, few retailers are allowed to stock their champagne, although many would like to do so. At least one imitator, however, has tried to pass off a common sparkling wine as a famous Krug. The foil on the genuine Krug bottle is of a brownish hue of gold, whereas the foil on the imitation is much brighter and more yellow. The labels are almost identical but are positioned differently. The genuine Krug carries its label at the bottom of the bottle, whilst the counterfeit is higher up. Only someone who knew Krug champagne very well indeed, would be able to tell the true from the false, at least from a cursory glance. When the bottles are opened, the words 'champagne' and 'Krug' appear on the cork of the genuine Krug but not on that of the false. Needless to say the wines bear no relation to each other.

When the fraudulent bottle was discovered, the police were informed and efforts were made to trace the source of supply. Unfortunately, like many similar offences, the rogues were never traced.

THE WINE LIST

The following story was published in Germany under the heading 'The Wine List'. A friend of mine who had kept the cutting for years gave it to me. The tragedy of the story is that it is true.

> In the olden days when the modern wine laws with their strict regulations were as yet unknown and vintners had rather a flexible conception of naming wines, a stranger once dropped into the little wine shop of a Bacharach vintner. He asked for the wine list and to his great amazement he saw that a vast choice of wines with high-sounding names were listed. After having taken a goodish time and a great deal of thought, he chose a bottle that promised to satisfy his fussy palate.
>
> This happened to be autumn. Whoever had hands that could be put to some use was out in the vineyards collecting the grapes. Only old granny, who as a rule did nothing but rest on a bench in front of the fireplace, had stayed behind. She was looking after the house. So the customer asked her to serve him. He shouted the name of the wine he had chosen into granny's ear; she nodded her old head and stepped shakily down into the cellar. So long a time passed without granny returning that the customer began to fear that something untoward had happened to the old woman. At last she reappeared in the room, hale and sound. In one hand she carried a bottle, in the other a cigar box. She placed both on the table in front of the stranger.
>
> 'There you are, sir', she said, 'There is the bottle of wine and the cigar box contains the labels. Now just find the label

with the name you fancy and stick it on the bottle yourself. My old eyes are no good for reading any more . . .'

Hirondelle Relabelled

This malpractice occurred in England. A well-established firm of wine merchants admitted relabelling 14,500 bottles of Hirondelle wine and selling it as Tyrolean Riesling. The firm was fined the maximum of £400, with £250 costs, after pleading guilty under the Trade Descriptions Act to falsely labelling the wine. The Managing Director was fined £300 personally after telling the Court that 'it was all a mistake', but he went on to say that 'relabelling was general practice in the Trade'. Hirondelle does in fact come from Tyrolean regions, mainly in Northern Italy, and is shipped and bottled by Hedges & Butler, who own the brand name. They sell the wine on to a great number of wine merchants for distribution. The relabelled bottles omitted the words 'Hedges & Butler' and replaced them with the name of the firm in question. It is these words that in fact contravene the Trade Descriptions Act because the company did not ship or bottle the wine.

South African Franconian

The following letter to the editor of a wine trade journal, from Mr Edmund Penning-Rowsell, a great expert and popular writer on wine, is self-explanatory and needs no further comment:

'After reading Colin Parnell's article in the September issue, in which he suggests, correctly in my view, that 'champagne perry' has a closer link with champagne than seems altogether right, my eye passed to a coloured advertisement on the opposite page for a bocksbeutel of Franconian wine.

'But no! In fact it was for a South African wine masquerading as such. The term is not too strong, for not only the shape *and* the colour of the bottle, but the name Grünberger Stein and the label, highly similar to the elaborate label with heraldic device, all suggest a Franconian source. And of course Würzburger Stein is not only the most famous of its vineyards, but is widely used to denote all Franconian wine; while there is a fair imitation of the German red wine seal, permitted only when a wine is superior to the general level of its grading.

'At such an example of passing-off the simple marketeers of Shepton Mallet would be likely to turn in their fermenting vats; and even the Mateus Rosé people didn't attempt to use German-type name, gothic lettering and Franconian-style label.

'The conclusion that many will draw from all this, is that if such imitation is necessary to sell the wine, then it cannot stand on its own and is unlikely to be much good.'

Good Wines in Poor Years

The German Spätlese and Auslese wines are very scarce in poor years when the demand still remains high but when there are few offers of good wine. A 'great' Mosel grower was known to succeed in these secondary vintages. He actually blended up his secondary vintages to Spätlese standard by chaptalisation and a share of good wine from a good vintage. Then he sold the secondary vintage as a good vintage wine. When his secondary vintage wines were tested, tasted and examined, the wines that everybody thought could not be Spätlesen were in fact Spätlesen at least to a large part! And it took a long time for the authorities to begin making investigations in earnest. They found what nobody could expect. He had for several years bought specially good Riesling Süssreserve not from his Mosel region but from the Rheinhessen, which gave his wines of secondary years the body to be presented as a good vintage wine. The case never came to Court and in my opinion it was due to his failing illness. So as not to damage a 'great' name, the hearing was postponed a few times until the man died. It was one of the most interesting cases, especially because nobody had ever expected it from him!

An Attractive Name

Piesporter Goldtröpfchen was one small vineyard in the village of Piesport on the Mosel. The name was attractive in all languages and the wine very acceptable. The demand soon became much greater than could be supplied. The vineyard was therefore expanded to the neighbouring villages, but of course still sold the wine under the name Piesporter Goldtröpfchen. After all this manipulation, one Trier merchant exported more wine under this label than had been harvested altogether.

There was no complaint. I am sure that the wines sold under this label were of good quality, perhaps equal and even better than the Goldtröpfchen vineyard wine. But because they were from lesser known vineyards and therefore cheaper for the merchant to buy, he could sell at prices well below his competitors, who bought the real Goldtröpfchen wine. They had perhaps to pay the same price for their wine as the fraudulent merchant charged for his. This unfair competition has been going on since wines were offered under single vineyard names, and this is generally only discovered on account of price cutting.

The Elastic Goldtröpfchen

Piesporter Goldtröpfchen includes in its 122 hectares of vineyard a number of neighbouring villages. The production in the vineyards of its district was an average of 9,300 litres per hectare for the vintages 1977–81. From 1978 to April 1983, 10.8 million litres were submitted for quality examination (6.86 million litres were Qualitätswein and 3.96 million litres were Qualitätswein mit Prädikat). This quantity contained süssreserve and wines of other origin in the Mosel-Saar-Ruwer region amounting to 25% of the total. Accordingly, the 10.8 million litres should have contained 7.65 million litres of wine from the actual Goldtröpfchen vineyard in Piesport. This figure is in fact 35% above the statistical average!

German authorities will not agree, but one report in the *Weinwirtschaft* of 5 April 1983 shows what is actually happening. A grower who owns 0.6 hectares of the Goldtröpfchen vineyard has sold in four years 119,000 litres of Piesporter Goldtröpfchen including 113,000 litres of Prädikatswein which were produced by the addition of 2.65 tons of liquid sugar! A masterpiece of fraud!

The Biter Bitten

In one of my cases in 1930 a wine wholesale merchant bottled 600 litres of wine in quarterbottles and sent them free of charge all over Germany to wholesalers outside the wine producing region. He followed up the samples to a few of the most important wine houses and booked an enormous number of orders, all wine from the same famous vineyard. A competitor, visiting his customers some time later, found this wine everywhere and knew that the original vineyard did not produce half the quantity this merchant had sold. He took as damages for unfair trading a part of the profit and he was satisfied. There were no other complaints, but I would hope that the culprit had learned his lesson!

Mosel 'Spätlesen'

The German Wine Law of 1971 was in dispute in 1975 following the poor harvest of 1974, especially in the Mosel-Saar-Ruwer region. Before 1971, the weaker wines produced in this region used to be blended with the less expensive wines from the Rheinland-Pfalz. This produced an acceptable and saleable wine that could still be called Mosel, since 75% of the blend came from that district and only 25% from the Pfalz. After 1971, it became illegal to blend quality wines from different regions and the arrangement had to cease - officially.

The quality of the 1974 vintage was poor throughout Germany, but to everyone's surprise a number of the small family growers in the Mosel region submitted more wines than ever for classification in the higher gradings of Spätlese and Auslese. The conclusion drawn by many, in and out of the trade, was that the young and dry wines had been secretly dosed with extra sugar to make them stronger and sweeter and so become eligible for the higher and more valuable classifications.

The *Sunday Telegraph* reported on 30 May 1976, that *Handelsblatt*, the respected German business newspaper, had stated that there was no doubt that 'a considerable amount of 1974 Mosel wines had won "Prädikatwein" attributes labels only after sugaring in the cellar'. *Wirtschaftswoche*, the German business weekly journal, commented, 'German wine, once world famous, has fallen into disrepute.'

As already explained, the law permits the addition of sugar to Tafelwein and even to Qualitätswein to make them acceptable and saleable, but it is illegal to add sugar to Prädikat wines such as Kabinett, Spätlese and Auslese, etc. The quantity of sugar that may be legally added to a Qualitätswein often makes it superior in quality to a Prädikat wine, and not only in a poor year, but the Prädikat wines always command a higher price. The temptation is obvious ...

A Qualitätswein wine, Mosel Riesling of 50° Oechsle (5.9% alcohol) in 1974 and a similar wine of 57° Oechsle (7% alcohol) in 1975 can be compared in the following way. In 1974 the alcohol content of the wine could be increased by 4.5% because it was such a bad season. In 1975, however, the alcohol content could only be increased by 3.5%, thus the 1974 Qualitätswein could become 10.4% alcohol and the 1975 10.5% alcohol. The minimum figures for Prädikat wine, however, are as follows:

Kabinett	*minimum*	70°	Oechsle -	9.1%	*alcohol*
Spätlese	,,	76°	Oechsle -	10%	,,
Auslese	,,	83°	Oechsle -	11.1%	,,

This shows clearly that enriched Qualitätswein has basically more alcohol and after the addition of Süssreserve; when cleverly selected with an especially good grape aroma, this will appear as a better wine than the Kabinett and Spätlese wine. This fact did not escape the Mosel growers and they fell *en masse* to the temptation: they classified their unripe and enriched Qualitätswein as a Spätlese. We now know also that 1974 was the vintage when they started to use liquid sugar for enrichment and as a substitute for süssreserve.

EASTERN FIDDLERS

Japan

Many people may not know that wine grapes have been grown in Japan for some years. By European standards the quality of the wine made from them still has some way to go, but is improving. Nevertheless, it is making a dent in the traditional market for Sake - the alcoholic beverage made from rice.

From a European point of view the scandal with Japanese wines is that they are made up to appear more of European than of Japanese origin. Bottles are mostly of an identical shape with those used for French Burgundy or Bordeaux wine and the labels display French names. To mislead the Japanese consumer still more, the labels sometimes show a château, or have a château name printed on them followed by the customary words 'mis en bouteille au château'. Names like Château Semillon, Château Lafife or Château Lafutte are used, also 'Estate Wines' or 'Estate Bottled but originating in Japan'. There is also an Ohasama Edelweiss on the market, as *Der Spiegel* reports. The consumer is assured, on the labels, that the Japanese wine is 100% Japanese, but the European wine scandal brought into the open that Japanese wines usually contain little, if any, more than 5% wines of Japanese origin!

The quantity of Austrian wine used in these blends was very

small. Most of the blending wines originate in Bulgaria and Yugoslavia where the duty for these wines is lower. The Japanese blenders cannot be said technically to adulterate the wine, as Japan has no wine laws, nor indeed labelling regulations. The outcry against the misuse of these 'lawfree zones', however, has forced some leaders of the Japanese trade to retire.

China

Grapes for wine have been grown in China for many centuries and in the last thirty years the industry has started to develop along modern lines. In September 1985 it was reported that three men had been executed in the Central China province of Sichuan for making and selling industrial alcohol disguised as rice wine which killed twenty-five people. It is clear that fiddlers in Chinese cellars will get short shrift.

Russia

It was announced in mid-October 1985 that Russia had become the third largest wine producing country in the world. Italy and France remain first and second respectively, whilst Spain has slipped into fourth place. There are fiddlers in the Russian cellars without doubt, but the Iron Curtain successfully hides them from the world.

17

The Algerian Argument

Before the outbreak of war in 1939, French wine was quite cheap in the United Kingdom. Australian wine was popular, especially the Emu brand; so, too, was wine from South Africa and Cyprus, but Algerian wine was virtually unsaleable at any price. Although Algeria was part of France and there was much to-ing and fro-ing of wine between them, the English wine drinkers would have none of it. The odd shipper who imported a few casks of Algerian red wine, almost certainly blended it with other wines and sold it by a fancy name. Algerian wine as such was seldom, if ever, listed by a wine merchant.

When France succumbed to the German invasion, many ships of the French merchant fleet were on the high seas loaded with French and Algerian wine. At the roll call, some 13,000 casks, each containing 660 litres of Algerian red wine, were brought to England. A veritable bonanza compared with the 30 casks that were imported during the course of a whole year before the war. Every bonded warehouse in Bristol, Cardiff, Manchester and London, became choc-a-bloc with Algerian wine.

Government officials contacted those companies still in existence who had previously traded with Algeria and offered to sell the wine to them, but to no avail. They were too few and too small and with inadequate facilities and staff. They suggested that the wine be poured down the drains! Victor Turner then advised the officials to contact the House of Hallgarten, well known for their importation of German wines and now without a source of supply. The officials consulted me and I examined and tasted the wine. My report was short but to the point. 'The wines are of different qualities - claret type, burgundy type and all good, but their name is bad.'

I made three proposals. First, that the wine be sent to Scotland and distilled into brandy. The officials would not even consider this. Second, that the wine be sold to the wine merchants at 3 shillings per gallon; a very low price even in those days for an unpopular wine, but it would have emptied the warehouses. Third, that the wine be sold as 'Red Wine, Produce of France.' It was good wine, but unsaleable as 'Algerian'. But as 'Produce of France' the wine would sell quickly. Algeria was, after all, part of France and the allegation was not so far-fetched.

The officials hesitated, debated and then sought other advice. The military authorities insisted that the warehouses must be emptied as soon as possible. The space was wanted for more important material as far as the war was concerned. I travelled from port to port; my brother got busy in London and a few people bought, but sales were slow. The Luftwaffe helped to solve the problem. During the air attack on London at the end of December 1940, the wine district around Mark Lane was blitzed and some bonded warehouses were destroyed. It was obvious that the losses would not be replaced, so wine merchants began to listen to us and soon the wines started to be dispersed to their cellars around the country, leaving the bonded warehouses free for other stock. Algerian wine was soon in demand and widely available - even in the best of London hotels. Some merchants made monthly allocations of their stock to their best customers until it was all gone, declining to sell it more quickly to all and sundry. Algerian wine with bottle-age achieved quite a name.

The importation of French wine had, of course, ceased. Stocks already in the country were diverted to the various officers' messes and clubs and soon there was no wine available for the man in the street and very little on the black market.

By May 1943, after the successful invasion of North Africa, a large stock of Algerian wine again became available and the Admiralty decided to bring it back to England in ships that would otherwise have returned empty. The person who made this decision presumably had no knowledge of the wine trade and the problems with Algerian wine. He no doubt thought that a wine-dry population would be glad to have some wine. There was no question of importation by the ordinary methods of trade and the military supply ships delivered the wine into the care of the Ministry of Food. The wines were fairly distributed by the Ministry to the previous importers of French wine, based on the proportion they imported in the three pre-war years. In time they got a very good volume of business, whereas pre-war importers from other lands looked on empty-handed. They all, first and foremost the former importers of German wines, pleaded for a share. Wine still arrived from Algeria but distribution was held up while the wine trade tried to devise a procedure for a fairer method of distribution.

While all the wrangling was going on, vested interests were at work to obtain imports of French wine in a matter of weeks after the invasion of France had been accomplished. By the time a decision on the procedure for distribution of the Algerian wine had been reached, French wines were again available. Those who had helped to delay the distribution of the Algerian wine no longer wanted their allocation and rumours went round that the wine had turned sour!

Bolder voices claimed that the Algerian wine had too much volatile acid and that it had turned to vinegar. The Ministry of Food was criticised for importing bad wine; there were questions in the House of Commons and, in 1946, even a debate on the subject. Sales of the wine had virtually ceased and anyone holding stock of it was anxious to get rid of it. My firm bought as much as was possible and encouraged others to do the same because the wine was not as it was described, with the exception of a few casks that had been on ullage too long. Nevertheless, sales remained very slow indeed.

Eventually, permission was obtained to export the wine to Germany where there was a great shortage of red wine. Some twenty-four expert buyers from different importers came from Germany and tasted more than 3,000 casks of wine. Only a very few turned out to have vinegarised, and these were promptly sold to pickle manufacturers. The rest of the wine was bought and shipped back to Germany at a price that just covered the expenses of the British Government. The rumours and criticisms suddenly ceased. The German buyers expressed their amazement that the English could refuse to drink such sound wine and went home feeling very pleased with their purchase.

Some months later, I was in Germany buying such white wine as was available. I enquired from all those merchants that I met; I looked through shop windows, but nowhere could I find any trace of the Algerian wine. Instead there was available for sale a remarkable number of bottles of red wine bearing pre-war printed labels of claret and burgundy. It became clear to me that the Algerian wine had been blended, or merely bottled and falsely labelled. In all probability some had come back to England and been sold much more expensively than the original Algerian wine.

The importers of French wine had been naturally anxious to protect their future sales of French wines. Had they continued to offer the cheap Algerian wine, their customers might have become accustomed to it and even preferred it to the more expensive French wine.

The calculated debasement of the Algerian wine was a scandal, especially as officials of the Ministry of Food had done everything possible to satisfy the claims of the trade for the import of wine. They were blamed without any justification; the wines were good on arrival, were well looked after and remained good. To safeguard their pre-war businesses, some importers deliberately and continually denigrated the Algerian wine until it again disappeared into obscurity. Algerian viticulture was virtually destroyed and with it the firm who built up the export to Britain.

For the consumer, this is yet another example of the need to evaluate wine objectively without being prejudiced by the label.

18

The 'King of Wines' is King Supreme

The cynic might well say that nobody protects him for his sake, what's in it for the trade? As in so many cynical remarks, there is an element of truth in it, but not the whole truth. Naturally, trade associations exist to protect the business of their members, but in so doing they also protect the individual consumer from being conned by interlopers. The famous 'Spanish Champagne' case exemplifies this very well.

THE HISTORY OF CHAMPAGNE

Britain has long been the major customer for champagne and for 150 years this exciting wine was well established in people's minds as the wine for all celebratory occasions. Ship building, a major industry, used vast quantities to launch their ships in festive style. Society balls and banquets, weddings and anniversaries, receptions and farewells, all are celebrated with champagne. The French champagne industry in the province of Champagne has always had a monopoly in supplying the wine for celebrations of every kind. Their customers have included emperors and kings of all the courts of Europe, heads of state, captains of industry and many others, too numerous to mention. The king of wines was the wine of kings and commoners alike.

Champagne is an unusual wine, quite different from all others. It is made in a north-eastern region of France where the sunshine is limited and the autumn soon turns cold. The grapes, therefore, are rather sharp, short of sugar and often poor in flavour. Furthermore, the onset of cold weather inhibits the yeast and fermentation of the grape must is therefore slow. When warm weather comes again in the spring the yeast often resumes the fermentation. For centuries this situation earned the wines of Champagne a poor reputation. They were virtually unsaleable.

At the end of the 18th century, however, Dom Perignon, the cellarmaster at the Abbey of Hautvilliers, bottled some wine during the winter, plugged the neck of each bottle with pieces of the new cork bark which had been but recently discovered, and tied them down with string. When spring came and the wine refermented, he opened the bottles and tasted the wine. The once harsh wine had 'miraculously' changed. It sparkled in the glass, was more fragrant, more robust and altogether more appealing than the still wine he had previously made. Unfortunately, the wine was somewhat cloudy since the second fermentation had created a further deposit of lees and this had been lifted up with the effervescence when the bottle was opened. Dom Perignon continued to improve the quality of his wine, notably by blending together grapes from different vineyards, especially those around the village of Ay, with those between Avize and Cramant. The poor, basic wine was steadily improved and the knowledge of blending was shared with others.

Many tried and failed to solve the problem of removing the sediment without losing the effervescence. Eventually, it was a woman who succeeded. Madame Clicquot was the young widow of a wine merchant who had determined to continue her husband's business. Her solution was to cut holes in a table and insert the bottles upside down. With a little shaking and twisting the sediment eventually slid down the side of the bottles and settled on the base of the cork. A cellarman then removed the cork whilst still holding the bottle almost upside down and the sediment came out with a rush of wine. The bottle was at once brought to an upright position and handed to a colleague who promptly topped it up with some similar wine that had been sweetened to improve the flavour. A new cork was quickly fitted and tied down. Little effervescence had been lost and the wine was not only sparkling but also star bright as well. Champagne as we now know it, had been born. Initially the losses from the burst bottles and the gushing wine were substantial. Heavier and stronger bottles were used together with better corks and wire cages. The losses, although reduced, continued at a worrying level until science came to the rescue.

The wine was kept sufficiently warm in the autumn by artificial means, for fermentation to be completed. It was then cleared, removed from its sediment and matured, blended with other wines and bottled. The secondary fermentation was created with the addition of a fresh yeast and a precise amount of sugar, calculated to produce a pressure in the bottle of 70 lb/in^2. The losses from bursting dropped dramatically and settled around 1 bottle in 200.

Losses from gushing wine steadily diminished, too, as the workers became more skilled. Today the bottle of wine is chilled hard and the neck is frozen. When the cork is removed the sediment encased in a plug of ice comes with it. Less wine is lost in the gush, less wine is needed for topping up. Other changes include the use of crown stoppers that so grip the neck of the bottle that wire cages are not required. Nevertheless, champagne is still an expensive wine to make because it needs so much handling of the individual bottles.

The whole process is known as the *méthode champenoise* to distinguish it from methods used elsewhere. These are the *cuvée close* method, where the wine is fermented in a huge stainless steel tank, then filtered and bottled under pressure; and the *carbonation* method, where the wine is not refermented at all but merely injected with carbon dioxide gas - the by-product of fermentation.

Champagne producers are reluctant to change the methods that they have laboriously evolved so satisfactorily. They are convinced that no other blend of grapes, method of vinification or method of re-fermentation in the bottle produces a wine as good as their 'real champagne'. They have successfully 'persuaded' other producers of sparkling wine, in Germany, Italy and France, not to call their wine champagne.

In Germany, sparkling wine is called *schaumwein* or *sekt* when it has nine months bottle age and Qualitätswein approval. In Italy, sparkling wine is called *Spumante* and is often made from the heavily perfumed muscat grape. In other parts of France sparkling wine is called *vin mousseux*. Only the wine from Champagne may legitimately be called champagne. The agreement now has legal enforcement throughout the European Economic Community. It was not always so.

THE BATTLE FOR THE NAME

In 1956 a small company called The Costa Brava Wine Co Ltd, was formed for the purpose of marketing in Britain a sparkling wine made at Perelada on the Costa Brava, in the north-eastern corner of Spain. It was a pleasant enough wine for the non-discerning, and had the merit of not commanding more than a modest price. Care was taken to ensure that the name did not infringe the copyright of other Trade Marks and the wine was called Perelada Spanish Champagne.

The establishment in the British wine trade quickly noticed the interloper, perhaps because they were already unwillingly tolerating a number of wines with the prefix Spanish, such as Spanish Sauternes and Spanish Chablis. These wines, although inferior to their French originals, were also cheaper and therefore cutting into their market. Spanish Champagne tilted the balance towards revolt and the wine was attacked in an address at one of the major wine tastings held in London in the summer of 1957. The Costa Brava Wine Co. Ltd picked up their cudgels and strongly retaliated through the wine trade press. Egotistically and perhaps somewhat foolishly, they suggested that the trade should either accept the name Perelada Spanish Champagne or fight it out in the courts. The gauntlet thus thrown down was quickly picked up by the Association of Champagne Producers. The latter was not alone, however; they were vigorously supported by INAO – the French Government-sponsored Institut National des Appellations d'Origine des Vins et Eaux-de-Vie. Champagne was not only rated their finest wine, but also earned a considerable amount of foreign exchange. Its sales must be protected.

Unfortunately, Britain had not then recognised the AOC (Appellation d'Origine Contrôlée) regulations which legally relate the name of a wine to the territory in which it is produced. Action could not therefore be taken by the Association under these laws, but two options were open. Criminal action could be brought under the Merchandise Marks Act of 1887, although there was the risk of an acquittal by an ill-informed jury. Alternatively a civil action could be brought under common law for 'passing off' as champagne a sparkling wine made in Spain. The risks in this action were incalculable. This was then new ground to our legal system and no lawyer could even guess at the probable outcome.

As in many activities, the greater the prize, the greater the risk. It would be wonderful to win the civil action and have the name champagne legally established in Britain as the exclusive description of sparkling wine made in Champagne in France. But fear of the calamity that would be caused by losing such an action, prompted the Association to opt for the lesser prize of a 'guilty' verdict under the Merchandise Marks Act of 1887. They rated their chances of success here a good deal higher, and most men would have made the same decision in their place.

The case eventually came before Mr Justice McNair on 17 December 1958, at The Old Bailey in London. Counsel for the Association, Mr Geoffrey Lawrence, QC, alleged that the word 'champagne' in Britain, meant *only* the sparkling wine made by the *méthode champenoise* in Champagne. He claimed that the phrase 'Spanish Champagne' implied that the wine was real champagne from Spain. This was a lie, the more so because Spanish Champagne was not re-fermented in bottle, but in a huge stainless steel or glass-lined tank, the *cuvée close* method, that did not produce such a good wine. Furthermore, it was misleading to suggest that ordinary people would possess sufficient knowledge to understand that Spanish Champagne was actually made in Spain and was not the real champagne from France. He made an extremely strong case with force and eloquence.

The Association of Champagne Importers saw Spanish Champagne not only as a threat to the quality trade as a whole, but also to the integrity of the wine trade as a whole. If the defendants were not found guilty, then anyone could pass off anything as something quite different and the reputation of the trade would disintegrate. Expert witnesses were called from the many aspects of the trade to support their claims.

Mr Gerald Gardiner, QC, the Counsel for the Costa Brava Wine Company, relied on custom and trade usage for his defence. He quoted numerous examples of geographic prefixes to generic names, such as Australian Burgundy, Chilean Barsac, Spanish Sauternes, etc. These names indicated the *type* of wine in the bottle and Spanish Champagne only did the same. It indicated that it was a champagne-type wine from Spain.

A growing section of the wine trade was involved in selling inexpensive wines to a developing market and, therefore, gladly supported The Costa Brava Wine Company. They also saw this as a test case by INAO. If the defendants were found guilty, the next step might be the recognition by the British Government of the Appellation d'Origine Contrôlée Regulations. This would undoubtedly be a threat to the downmarket trade, and an end to the blending of ordinary wine with AOC wines whilst retaining the AOC name. There was no shortage of expert witnesses to support the defence.

The case lasted six days and then the judge summed up, seemingly favouring the prosecution. He told the jury to consider the meaning of the word 'champagne' in Britain and warned them not to accept the argument that the prefix 'Spanish' made it perfectly clear that the wine was not French champagne. He said that one could not take a noun with a well known meaning (i.e. champagne) and add to it an adjective (i.e. Spanish) inconsistent with that meaning and then say, 'Well, taking the two together, they are not false, they are true.' The jury, however, took less than one hour to find the defendants not guilty of any offence under the Merchandise Marks Act. Mr Justice McNair, seemingly supporting his summing up, said that it was a reasonable and proper action to bring and worthy of the court hearing. Nevertheless, he ordered the prosecution to pay the legal expenses of the defence!

The Costa Brava Wine Company lost no time in making the most of the publicity of their acquittal. Sales continued to boom. The champagne interest, however, withdrew for re-grouping.

After a surprisingly short period of consideration and with highly commendable courage, the twelve most famous champagne houses exporting to Britain, joined together in a consortium to take civil action against the Costa Brava Wine Company, to protect the good name of their sparkling wine. They sought to obtain in the High Court an injunction forbidding that Company from 'passing off' a sparkling wine as champagne which was not champagne.

The Costa Brava Wine Company was understandably aghast. Having been acquitted of any wrong under the Merchandise Marks law enacted by Parliament, there was no known precedent in English Law for this kind of action. Whilst an individual could take action against another to protect his good name, it was unknown in Britain for a group of people to take action against an individual to protect the geographical name enjoyed by their product. The Company asked for the action against them to be dismissed, without the necessity for expensive legal representation at a long Court hearing.

The Court, in fairness to the Costa Brava Wine Company, agreed to consider this point of law first. Both sides engaged eminent counsel and began the argument before Mr Justice

Danckwerts in November 1959. Counsel for the plaintiffs, Mr Richard Wilberforce, argued in minute detail for two whole days, that English Common Law did not preclude the protection of a geographical name. Sir Milner Holland, for the defence, argued with equal fervour, attempting to discredit his opponent's arguments and develop his own thesis that there was no relevant point of law upon which an action could be brought.

Mr Justice Danckwerts listened carefully, and finally decided that the plaintiffs did have a right in Law to protect the name 'champagne' if they so wished. But this was only the point of law and not the specific case itself. The fight was not yet won, indeed had not yet begun.

The consortium of champagne producers now had to prove that someone might buy a bottle of Spanish Champagne believing it to be real champagne and that this had an exclusive and non-generic meaning. It took them a whole year to develop the details of their case which they presented, again before Mr Justice Danckwerts, on 29 November 1960.

Mr Geoffrey Lawrence, for the plaintiffs, presented twenty-one witnesses, but the witnesses for the defence had shrunk to only four. The two major factors in favour of the plaintiffs, however, were a printed menu card mentioning 'Champagne (Perelada)', and a pink and blue brochure published by The Costa Brava Wine Company, entitled 'Giving a Champagne Party'. In both instances, declared Mr Lawrence, there was evidence of intention to deceive the public into thinking that they were drinking real champagne. The Judge agreed that 'On the face of it, it is quite plainly intended to cash in on the reputation of champagne.' Sir Milner Holland, again the Counsel for the defendants, spoke for two whole days, developing the points made in his cross-examination of witnesses. He argued that very few, if any, members of the public would be deceived and certainly not the 'substantial section' claimed by the plaintiffs. Mr Justice Danckwerts listened in silence and when the case ended, he took another two weeks to compose his judgement. It was not until 16 December, just two years from the date of the first action, that he read out to the assembled lawyers, their principals, the press and the public, the decision which he had so carefully considered. The Judge found in favour of the consortium of champagne houses. He granted them an injunction restraining The Costa Brava Wine Company from calling their Perelada, or any other wine, any name that included the word 'champagne' on the labels of all their existing bottles of sparkling wine with effect from 'within forty-eight hours'. There was no appeal.

Thus the wine trade, first through the initiative of the Association of Champagne Importers and then through the determination of the twelve champagne producers, successfully defended the name of their wine. At the same time, they safeguarded the British public from having any kind of sparkling wine passed off on them as champagne.

The outcome of course did not change the fact that in most years the basis of champagne is undrinkable 'Champagne nature'. Furthermore, champagne may have a better name, it need not have a better taste. What one likes best can only be discovered in a blind tasting. I once put ten sparkling wines, among them one bottle of champagne, before a dozen young wine consultants to find out the best and the champagne.

Whatever they found the best they considered to be the champagne among the sparkling wines presented. Far from it; most of them did not find the champagne, and considered sparkling wine from Luxembourg, Burgundy and Israel to be better! And what a saving that could mean....

The Champagne Houses arranged a champagne tasting at the International Convention of the Food and Wine Society in Torquay. By seeing and tasting a dozen champagnes side by side and, mind you, champagne from the great Houses only, one could see the difference, but there would have been space for good and better sparkling wine in between.

19

Other Sparkling Wines

THE BABYCHAM CASE: A BRITISH ARGUMENT

More recently, in 1975, a somewhat similar case was brought against the makers of 'Champagne Cider' and 'Perry Champagne'. Again, judgement was given forbidding the use of the word 'champagne' with any descriptive or qualifying word at all. Champagne is unique!

But Showerings, a subsidiary of Allied Breweries, who produce 'Babycham' from pears, would not give in and subsequently appealed. In the summer of 1977, by a majority of two to one, the three judges in the Court of Appeal reversed the decision. Lord Justice Buckley and Lord Justice Goff pointed out that the champagne industry had not been able to produce a single witness who had been confused by the Babycham label. Lord Justice Waller, however, dissented. He felt that the possibility of confusion among 'young, female, inexperienced consumers must be great'.

M. Joseph Dargent, the head of the Department of Information of the Comité Interprofessionel du Vin de Champagne, who had brought the case against Showerings, declared at a press conference following the judgement: 'We have lost this battle but not the war.' He felt that the CIVC had some good grounds for appeal to the House of Lords or even to the European Court in Luxembourg.

M. Dargent explained why he thought that it was so important to protect the name champagne. 'There is a certain amount of snobbery attached. We have been producing top class wines in the Champagne region for more than 300 years and we do not want its good name to be debased by beverages of lower quality. If all the beverages were called champagne, the word would lose its glamour and image.'

At a party organised to herald another victory for champagne, but which served to drown the sorrows, M. Dargent was handed a glass of Babycham. With all the panache of the connoisseur that he is, M. Dargent looked at the bubbles and conceded: 'It looks like champagne', then he nosed the bouquet and promptly exclaimed, 'but it smells like pears'. The throng of people around him gave him no option but to taste the beverage. With the courage of an army officer he took a sip, stepped back a pace and exclaimed, 'It's much too sweet'. He declined to comment further and never mentioned the dreaded name Babycham!

It is very flattering for Babycham that the Champagne Houses feared the competition of a product that costs but a fraction of theirs. After all, the unripe grapes used for making champagne are full of malic acid; is there a real danger that a person tasting Babycham may consider it to be real champagne?

OTHER COUNTRIES

In some other wine producing countries, the name 'champagne' is still applied to sparkling wines and M. Dargent has plenty of work left to do, before the whole world recognises the protected name.

Australia

Many wineries in Australia make a sparkling wine that is marketed as 'champagne' solely for those wines made exclusively by the *méthode champenoise*. Mostly the *cuve close* method is used or a new variation whereby the wine is re-fermented in bottle and then transferred to a large, sealed tank for fining, sterilising and sweetening before being bottled under pressure. The resulting wine is often extremely attractive and said to be the next best thing to champagne. Of course the Australians know full well that their wine is made in Australia and no one believes that their 'champagne' was made in France.

Canada

Here a high proportion of the people adhere to the French culture, particularly in Quebec, and the government recognised the exclusivity of the name champagne as far back as 1933. Consequently, it was made illegal to attempt to sell in Canada any wine bearing the name champagne, except that produced in Champagne and exported to Canada. The French Canadians love their champagne and currently import some 2,000,000 bottles each year to celebrate their happy events.

At the beginning of 1978, however, a new champagne war started between France and Canada, not about the name of the wine but for political reasons. Towards the end of 1977, the Prime Minister of Quebec, who had been leading the movement for separation from Canada, was received in France with honours that are normally reserved for the Head of a Sovereign State. Furthermore, France indicated that she supported Quebec in her efforts to become a separate State. By way of retaliation, the Canadian Government gave notice of the termination of the 1933 agreement over champagne. Canadian manufacturers of sparkling wines received the news with enthusiasm and jubilation. The Quebec Government countered that it would continue to honour the agreement and that it would not permit the sale of any sparkling wine names as champagne unless it had been imported from Champagne. This, the Government claimed, would protect the residents of Quebec from having passed off on them as champagne, wine that was not champagne. And so the squabble continues, making more work for poor M. Dargent.

United States

On the other side of the coin, Möet and Chandon, who supply one in every three bottles of the champagne consumed in the USA, moved into the Californian wine trade in 1977. They acquired 1,280 acres of vineyards in the well known Napa Valley and built a modern winery to produce a sparkling wine as near to champagne as any wine not produced in Champagne can be. They imported the champagne yeast from the Champagne estate and skilled workers to teach the American workers the tricks of the trade, especially that of *remuage*. The wine is to be labelled Domaine Chandon, and the association of this name with the Möet and Chandon reputation for fine champagne is clearly a winner.

SPUMANTE

A giant British conglomerate lost no time in avoiding trouble in 1976. It imported muscat grape juice into its winery, made a sparkling wine from it and marketed it under the name Cascato Spumante. The wine was launched in a gale of publicity on 22 April 1976. Their retailers and competitors promptly questioned the firm about the legality of describing a non-Italian wine as a *spumante* - the traditional Italian name for a sparkling wine. They took the point and promptly decided that discretion was the better part of valour. The label was redesigned immediately omitting the word spumante and inserting the words 'Sparkling British Wine'. Few confusions can have been so quickly corrected. The wrong labels lasted only a fortnight but even so, 500 dozen bottles were sold. That means that as many as 6,000 buyers could have been fooled into thinking they were buying a bottle of real Italian spumante when all they were getting was a British sparkling wine - no matter how good the comparison. Labels are the *bêtes noires* of oenophiles. Wine lovers will drink almost any wine, albeit some with more pleasure than others, but they do like to know just what they are drinking.

SEKT

A famous actor of the 19th century, Ludwig Devrient, had a great liking for champagne and this was the only wine he ever drank. In 1815 he was playing the part of Falstaff in Shakespeare's play *Henry IV* at the Berlin Court Theatre. In the play, Falstaff constantly called for 'sack'. When roistering after the performance, Devrient would call out for a glass of 'sack' which he pronounced 'sekt'. The intelligent barman promptly served him with the champagne that he knew Ludwig wanted. His companions imitated their hero and soon the word sekt became the fashionable name for champagne in German court circles where it was particularly popular.

One of the articles of the Treaty of Versailles prohibited the use of the name champagne in Germany, and so in the early 1920s the German wine trade officially adopted the name sekt to describe their wine, so similar in style and character to champagne. A lesser quality sparkling wine is known as *schaumwein*. It is nearly always a blend of imported wines as well as indigenous wines not suitable for sale as still table wines.

Because Germany produces mostly white wine, due to her northerly situation, the industry has developed advanced techniques in white wine production. With this knowledge, producers of sparkling wine in Germany are able to select the most suitable wines for blending as the basis for sekt, not only from their own country but also from wine producing countries throughout the world.

For nearly sixty years these wines have been known as Deutscher Sekt and the Wine Laws of 1971 gave confirmation to this. They established that the name sekt should be reserved for a quality sparkling wine produced in Germany and that a new category be established to be known as 'Prädikatsekt'. This name was to be reserved for those quality sparkling wines that consisted of at least 60% German wines and not more than 40% imported wines. Both sekt and Prädikatsekt wines had to pass the official tests of authenticity in either category before a certificate and Prufungsnummer could be granted. This equated sparkling wines with the three categories of still wines: schaumwein equal to Tafelwein, sekt equal to Qualitätswein and Prädikatsekt equal to Prädikatswein. But the basis for all three is of course undrinkable Tafelwein - only a fool would buy anything else for the manufacturing of a sparkling wine.

More recently, however, sekt has been a cause of concern to the EEC. Its definition as German sparkling wine has been under consideration by the Court of Justice of the European Communities in Luxembourg. The decision of the Court, legally binding on its member countries, caused surprise and disappointment in the German wine trade. They had long grown accustomed to the name sekt being exclusively reserved for German sparkling wine. They had always considered the name as a designation of origin linked with the place of production, i.e. Germany.

The Court held, however, that the name sekt did not indicate any specific source of origin but was rather an indication of the wine's composition. In their view the designation sekt was but a general description for quality sparkling wines and the title was not exclusively German and could be used by any member country.

German Prädikatsekt with 60% German wine as a basis has also become a wine of the past. The European Court found this regulation to be not in accordance with the EEC. Agreement. The

German sparkling wine is made from wines imported from the South of France, the Loire and wherever undrinkable wine is available to be put through the sparkling process.

Before the law suit started, the German sekt industry discussed with the Asti producers and the Champagne Association the possibility of a compromise. They proposed that in Germany champagne could be sold without the addition of the description schaumwein. The same could apply to Asti. In return, both France and Italy were to allow Germany the exclusive use of the name sekt.

The compromise failed because of the intervention of the French producers of *vin-mousseux*. The producers of this category of sparkling wine, for ever relegated to second place in France because of champagne, could see a great opportunity to get into the German market with the aid of the name sekt.

Adversity being the mother of invention, however, the German wine trade is now considering the use of regional description, such as Rheinsekt, Moselsekt and Saarsekt, which are well within the law and yet even more indicative of the German origin of the wine. Lately, Lagensekt and Vintage Sekt is offered by some growers, co-operatives and even estates. The first of these was the Schloss Johannisberg sekt, which after growing demand became the brand 'Furst Metternich' - *Bereich* Johannisberg, which need not include a drop of Schloss Johannisberg!

20

Mr Justice Cross Defines Sherry

The decision in the 'Spanish Champagne' case in 1960 reverberated around the British wine trade, bringing fear and trembling to firms at risk. Three such firms were: Vine Products Ltd, Whiteways Cider Co. Ltd, and Jules Duval and Beaufoys Ltd. For many years these firms had been importing concentrated grape juices - mostly from Cyprus and Greece. The thick juices were then diluted with water, blended and fermented with yeast. The resulting wines were marketed as 'British Wines'. They were inexpensive and very popular with certain people. Millions of bottles were sold each year. Among the most successful wine styles were a number of sherry-type wines.

Now sherry, or at least wine from Jerez de la Frontera, had been imported into Britain since long before the reign of the first Queen Elizabeth. Indeed, there is a manuscript in a Spanish library referring to the export to Britain, as far back as 1561, of '4,000 butts annually' (2,592,000 bottles). Shakespeare immortalised the name 'sherris sack' through the words uttered by Falstaff in his play *Henry IV*, *Part Two*.

The popularity of sherry rose and fell, reaching a peak in 1870, then falling away before slowly climbing back, until in 1965 some 5½ million gallons (33 million bottles) were imported from Spain. The wine was always called sherry and in 1935 the Spanish Government legalised this long standing practice by authorising the producers in the Jerez district to call their wine 'Sherry' or 'Xeres' or 'Jerez', all variations of the same word and deriving from the name of the town in the heart of the area producing the wine. The words 'Spanish Sherry' had never been used any more than the words 'French Champagne' in the previous case.

The history of 'British Wine' is almost as long. In 1635 the English Government granted an exclusive right to Francis Chamberlayne to make and sell wine produced from 'dried grapes or raysons'. Later, other ingredients were included with the raisins, particularly rhubarb, although cherries, gooseberries, blackcurrants and similar fruits were also used. Apart from being sold direct to the public, these British wines were also sold to the merchants in the wine trade, where they were blended with imported wines that had been knocked about by storms whilst coming from France in the small sailing boats used for that purpose. The wines were then sold as 'best quality'.

By the middle of the 19th century, 1852, some 600,000 gallons of British wines were being made and sold every year. Fifty years later a method of dehydrating grape juice was discovered and opened up new possibilities in the manufacture of wine in places far from where the grapes were grown. By 1965 in excess of 9 million gallons (54 million bottles) of British wines were made commercially from these concentrated grape juices.

The manufacturers of these wines, however, rarely described their wines simply as 'sherry', but nearly always added the prefix 'British' or included words explaining that the wine was made in England from the juice of foreign grapes.

In 1925, the Sherry Shippers' Association brought an action under the Merchandise Marks Act of 1887 against a firm selling 'Corona Pale Sherry' - produced in England from the juice of selected foreign grapes. During the hearing a label for 'Rohilla British Sherry' was shown to the court. Counsel for the Sherry Shippers' Association commented: 'Yes, that seems to comply with the proviso.' By that he meant that the expression 'British Sherry' was established prior to the passing of the Act in 1887, and was accepted at that time as meaning a British made wine of sherry character. The Sherry Shippers' Association never prosecuted if the title 'British Sherry' was used, although they were invited to do so on a number of occasions by the Magistrates. Subsequently, 'Australian Sherry', 'South African Sherry' and 'Cyprus Sherry', were imported and sold as such without objection from the Sherry Shippers. During all this time, sherry from Spain was imported and sold under the simple name 'sherry'.

Because of the decision in the 'Spanish Champagne' case, however, the three companies already mentioned feared that similar action might be taken against them. On the basis that the best defence is attack, the companies, all members of the Showering Group, brought a civil action against the Big Four Sherry Shippers - Mackenzie and Co, Williams and Humbert Ltd, Gonzalez Byass & Co, and Pedro Domecque S.A. - to obtain a declaration that the words 'British Sherry' did not infringe their rights. The four shippers were actively supported by their Association and other producers. The plaintiffs called many witnesses, displayed ancient wine lists, quoted previous cases. Both sides produced immense detail to support their claims and counter-claims. An important principle was at stake.

The High Court hearing lasted twenty-nine days and it was not until 31 July 1967 that Mr Justice Cross read his detailed judgement. It took him eighty minutes. In essence, he declared that the word 'sherry' by itself meant wine of a particular characteristic, produced exclusively in the Jerez district of Spain. Nevertheless, because the title 'British Sherry' had such long usage – at least one hundred years – it could continue to be used.

He pointed out that sherry producers and shippers had known of this name all along and that it was, in fact 'common knowledge in "Sherryland"'. They had by long acquiescence lost their right to complain about wines of a sherry-type that did not come from Spain and were described as 'British Sherry' etc. Nevertheless, they had shown by frequent litigation that they had not acquiesced in such wines being called simply 'sherry'. Henceforth, he declared that all sherry-type wines produced other than in the Jerez district of Spain must be labelled with a geographical prefix before the word 'sherry'.

'British Sherry' is such an established designation that no consumer can be misled, and it was hoped that this question had been buried for all time. That is not so, however; with the entry of Spain into the EEC some vested interests wish to have the word sherry reserved exclusively for Spanish sherry and the first steps seem to have been taken in this direction.

It would be regrettable if acquired rights confirmed by the High Court could be undermined and destroyed by the European Community, be it by legislation or jurisdiction. If the protection of the consumer is the purpose, I think that the English Courts know the mentality of the British consumer better than any other Court.

It has been noted that in recent years a number of suppliers, especially of 'own brand' sherries, specify their wines as 'Spanish Sherry' to emphasise their authenticity as real sherry.

Alerted by their success with analysis of Port Wine, the West German analysts also applied the C14 test to some casks of sherry. Mr John Morris, the Secretary of the Sherry Shippers Association said that the reports that reached him in April 1975 were controversial. He continued, 'I understand that certain casks from one consignment of sherry have been tested in Germany and some samples were found to be over a mean line, while others from the same batch were below it. This, I would say, means that the analysis method is subject to a fairly wide range of inaccuracies. It could be dangerous to draw any hard and fast conclusions from these tests.'

We can breathe again, confident that the watchdogs are awake and alert. The problem of finding out where the responsibility might lie in Spain would be far more difficult than in Portugal. In Spain the producers of sherry are legally free to buy their grape spirit to fortify their wine from whom they please. Very many delicate enquiries would have been necessary to find the source of an illegal spirit. It would be extremely difficult to persuade anyone to disclose this kind of information. On the other hand, once it is known that 'the cat is out of the bag', producers often take fright and revert to orthodox methods.

21

The Montilla Case

Not very far from Jerez de la Frontera lies the ancient city of Cordoba. It is the centre for six wine producing communes, the best known in England being Montilla. Wines very similar in style to sherry are produced, although they are not fortified. The wines are fermented to produce as much as 14 or 15% alcohol and then left, often in the open, in giant jars holding 5,000 litres each, to develop their distinguishing characteristics – *fino*, *amontillado* or *oloroso*. Originally the word amontillado was applied to a wine from Jerez to indicate that it was like a wine made in Montilla. Now it has come to mean a medium dry wine in the sherry style.

Montilla wines were not very well known in the United Kingdom until the last twenty years or so. Then the cost of sherry began to rise because of the increase in Excise Duty on fortified wines. Because Montilla wines are not fortified, the duty is somewhat lower, enabling the wine to be sold for less than sherry.

All might have been well had the Montilla importers left the wine to sell under its own name. But no doubt to differentiate between the different styles and more especially to promote the sales of the wines by indicating their similarity to sherry, they described the Montilla wines as being fino, amontillado or oloroso – descriptions that had previously applied in Britain only to sherry.

The sherry producers and shippers claimed that this was a deliberate attempt to pass off Montilla wines as sherry. Accordingly, Sandeman Hermanos, A. R. Valdespina and Duff Gordon, issued a writ against Western Licence Supplies of Bristol, in the British courts in 1970. They asked for an injunction against Western Licence supplies, prohibiting them from using the adjectives fino, amontillado and oloroso in describing the Montilla wines that they imported and distributed in the United Kingdom.

The legal struggle dragged on for nearly six years and was finally settled in the High Court by an agreement reached between the lawyers outside the Court. Western Licence Supplies, who command 80% of the Montilla trade in the United Kingdom, agreed to give up the use of the descriptions after 31 December 1976, by which time the existing stock would have been sold. They further agreed to pay £10,000 towards the plaintiffs' costs which were estimated to amount to more than £50,000. The precise judgement was given in the following terms:

'This court doth order that the defendants and each of them, be restrained (whether by their respective directors, officers, servants or agents or any of them or otherwise whosoever) from using in the course of trade the words "amontillado", "oloroso", or "fino" or any of them in conjunction with any wine not being wine coming from the Jerez district of Spain.'

Oddly enough, Montilla wines are legitimately described as fino, amontillado and oloroso in Spain. There the words are regarded as descriptions of styles of wine just as we might use the terms dry, medium and sweet. The Spaniards see no problem.

22

The Bordeaux Affair

On 18 December 1974, the presiding judge announced the verdict of the court in Bordeaux on eighteen men accused of having adulterated or falsely labelled at least three million litres of wine. The trial had begun early in the previous October and was scheduled to last no more than three days. In the event, such widespread allegations of fraud were revealed, that two whole weeks were needed to unravel the truth.

WORLDWIDE INTEREST

Bordeaux has for centuries been the capital of the wine world. Its name has become synonymous with good red wine. The prices obtained for red Bordeaux set a standard for red wine everywhere. It is commonly assumed that if the vintage is good in Bordeaux, that it affects the thinking of wine traders and drinkers the world over.

News of the trial for fraud in Bordeaux, then, hit the whole of the wine world like a bombshell. Interest was intense. At least two hundred journalists from all over the world attended the trial. The interest in the scandal was not confined to those working in the wine trade, but extended to everyone who had ever bought a bottle of Bordeaux wine, because in the dock stood two members of the family of one of the most famous shipping houses in the whole of Bordeaux.

THE PRINCIPAL DEFENDANTS

The firm involved was established more than two hundred years ago. An important Germanic colony developed in Bordeaux during the eighteenth century – the emigrés coming mainly from Hambourg, Lübeck and Hanover. The original head of the firm was one of these and, like others, he prospered as a buyer and seller of wines because of the lucrative alliances with merchants in Germany.

In the generations that followed, palatial houses were built and elegantly furnished. Their wealth made these families the leaders of Bordeaux society, looked-up to by all and visited by the world's well-to-do in search of good red wine. They travelled, and talked and tasted wines, pontificating their views and accepting, as of right, positions of authority and international recognition. Of them all, the family of the firm concerned were among the most prominent. News of *their* indictment, not only for fraud, but also for destroying incriminating evidence, shook the wine-drinking world to its foundations and scandalised people everywhere.

Three members of the family were accused, Herman, Lionel and Yvan. Only Lionel and Yvan stood in the dock, however. Herman died before the proceedings began. Also in the dock were sixteen others, including brokers, their assistants and the director of a Bordeaux oenological laboratory. The outstanding personality of this group, however, was Pierre, the wine broker who allegedly conceived the fraud and organised the chain along which 'inferior' wine travelled from the grower and became 'superior' wine when it reached the retailer.

Pierre came from 'the other side' of Bordeaux. He lived among a group of people called the 'Chartrons'. Originally these families had come from Ireland, Scotland and Prussia, but they had been ejected by the wealthy establishment and excluded from their neighbourhood. They settled near the river in a misty, marshy area once a Chartreux monastery, hence the description, Chartrons.

They became a separate, indeed almost a secret society, interrelated and inbred. Pierre himself was born in Barsac, one of the sons of a struggling vigneron, not always able to make ends meet. The grandfather eventually sold up and moved to Bordeaux where he formed a small firm buying wine from individual growers and selling it to traders in Belgium and Germany as well as in France. During the war of 1939–45, the firm continued to trade with its customers, including those in the Rhineland where there was a ready market for red wine.

After the war, Pierre's father was arrested, convicted of being an 'economic collaborator' and imprisoned. He died within a couple of years. Pierre had taken his father's place in the firm and when his grandfather also died, he was left to solve all the problems alone. Although he had inherited a substantial sales figure with newly developed exports to the United States, Canada, Great Britain and Scandinavia, he had also inherited some substantial debts, including a fine of £15,000 for the 'benefits of the war'. The young Pierre, it was alleged, soon learned how to 'stretch', 'doctor' and 'improve wines' to his advantage and how to 'cook his books'. By 1974 he was a well-spoken member of the middle class, with an extrovert and ebullient personality, comfortably at his ease with prosecuting counsel and judge alike. Almost at once, he became the 'star' of the case, the broker who had personally made £360,000 from the fraud. For when he was accused by the prosecutor he immediately admitted his guilt. 'I am guilty . . .'

But his admission was completely devoid of humility or any sense of shame; it included no awareness of wrong-doing or regret. Instead, he vehemently attacked the Appellation d'Origine Contrôlée laws and lashed out at the ineptitude of the men whose duty it was to enforce them. He aggressively asserted that much of the minimum quality wine entitled to the Appellation 'Bordeaux' was so poor, that fraud was necessary to make it agreeably drinkable and, therefore, saleable. He argued with an air of confidence and authority that out of 1.5 million hectolitres of wine entitled to the appellation 'Bordeaux', no more than half were worthy of that great name and the rest were so modest that they could not be offered to a foreign clientele. A great many people with knowledge of the Bordeaux trade privately agreed with him, but hastened to add that this was no justification for fraud.

Bert went on, 'There are two markets for the "Bordeaux". For the second there must be a mixture. The mixers are the growers and the sellers, and they sometimes make a very good mixture.'

WHY?

The background to this scandal seems to the layman to be equally scandalous. Why was it that the fraud succeeded for so long and how was it discovered in the end? Fraud Squad Inspector Roger Destrau admitted in court that the information for the raid had indeed come from a confidential 'adviser'. The French newspapers quickly pointed out that this was a standard euphemism for informer in the wine trade.

According to the Press reports of the hearing, Pierre said, 'I asked myself why on this occasion the inspectors did not ring up, as they had in the past, to say that they had some information. The matter would then have been settled with a handshake and a few thousand francs across the table ... After all, everyone had been doing it in one way or another for years and the inspectors had learned to live with the facts of life.'

Inspector Destrau told the crowded court that the enquiry started by chance on 20 June 1973, when the authorities received information that document record books were being fraudulently altered in the wine trade and that Pierre was involved.

The inspector said that he went without warning to the cellars concerned and Pierre reluctantly handed over his trading books.

'We went through the contents of the cellars, hour by hour, load by load,' said the inspector, 'and we found that all the documents had been falsified.'

'STEADYING A TOTTERING STRUCTURE'

The *Financial Times* headline just quoted, highlights the economic conditions at the time of the frauds.

One of the accused, President-Directeur of the Office de Vente des Liquides, told the court, 'Since 1972 we saw mad things. Because many Anglo-Saxons had monetary problems, they became interested in Bordeaux wine which had become like gold bullion. I have even seen dealers in rubber and cigarettes proposing to be my agent.'

The relatively minor fluctuations in stock and prices over the years before 1970 were of such a scale that the 250 *négociants* in Bordeaux could usually accommodate them without fear of bankruptcy. In 1970, however, the Bordeaux grape harvest was plentiful and of such good quality that prices went through the roof and a tonneau of wine (900 litres) cost as much as 4,000 Fr. The high prices of the 1971 and 1972 vintages were driven up by speculators and in four years the price of claret quadrupled. In 1973 there was a record harvest coupled with a world recession and prices tumbled. In the spring of 1974, the growers tried to obtain an agreed price of 2,000 Fr a tonneau for their 1973 vintage. They settled for 1,800 Fr, but many had to sell much lower, sometimes for as little as 1,000 Fr. Unfortunately, the over-generously awarded Appellations after the 1973 harvest had resulted in some very poor wines entitled to the certificate of origin 'Bordeaux'.

The 1974 harvest was also large and by the end of the year it was estimated that there was a surplus stock of red Bordeaux Appellation d'Origine Contrôlée wine of around 7 million hectolitres equal to nearly 1,000 million bottles, compared with a 2.5 million hectolitre surplus in 1969. The growers were holding stocks of wine that they could not sell to the *négociants*, who had stocks of wine that they could not sell to the shippers, who had stocks of wine that they could not sell to the retailers. All had paid very high prices for the wines of 1971 and 1972. Certain retailers were unloading wine at almost any price to cut their losses.

Many fine wines were sold at extremely modest prices for those who could afford to pay for them. Unfortunately the recession had caused a substantial downturn in the demand for wine. In the USA, Bordeaux AOC wines that had been fetching $11 a bottle for the 1970 vintage, fell to 4.99. In Japan, where wine consumption had been doubling year after year, 1971 and 1972 vintage Bordeaux AOC wines were offered in Toyko's most prestigious store for as little as £1.04 per bottle. By contrast a bottle of Romanée-Conti 1937 was priced at £160. It was estimated at the time that stocks of wine already in Japan would last for at least one year and possibly longer.

The Bordeaux market is so fragmented with countless growers, blenders, brokers, shippers, wholesalers and finally retailers - nearly all of them very small, albeit highly respectable - that rationalisation is urgently needed. Only large firms controlling all their sources of supply and distribution - as do the oil companies - are likely to have the financial resources to meet any new fluctuations in the future. The great need is for stability.

The layman, naturally, tends to push the economics of the wine trade out of his mind, to leave it to the experts. What he wants is good wine in his glass at a price he can afford.

'THE BETTER CLARETS OFTEN CONTAIN A BIT OF ALGERIAN'

Pierre continued, 'It can happen that the doctoring is good, but one indulges in a number of operations in which it is difficult to keep to the law. In the thirty years I have been at my job, I have seen fraud perpetrated everywhere, at the growers and in the trade. There are certain laboratories whose job is to improve bad Bordeaux wine and to make it better. The role of certain dealers is to collect bad Bordeaux wines and to improve them.'

'Had he sometimes mixed white wine with red?' he was asked by the prosecutor. 'Yes, it has happened. A little white does no harm to quality when there is too much tannin in the red.'

'But it is not legal,' exploded the prosecutor.

'No. But it is good,' asserted Pierre.

'But could it not be recognised in the tasting?' enquired the President of the Court.

'I leave that to the experts to decide,' responded Pierre. 'During all the time the fraud went on I never received a complaint from a client on the quality.' He estimated that he alone had fraudulently sold nearly two million bottles of wine, of which 75% were sold worldwide under labels marked simply 'Bordeaux', the rest under fancier labels suggesting a region such as 'St Julien, Médoc or Pomerol. 'I never harmed the reputation of a great wine,' he claimed.

THE GREEN DOCKET

His switching was restricted to the 'little Bordeaux wines', the humblest member of the wines in the Appellation category, known simply as 'Red Bordeaux'.

'And what is a Red Bordeaux?' Pierre asked rhetorically. 'An *honest*, saleable little wine that travels with a green administrative docket.' The reference was to the document switching operation that enabled a wine of 'inferior' quality or without an Appellation to be passed off as one with an Appellation and therefore of greater value. 'Were these ordinary wines good enough to be mistaken for Red Bordeaux?' asked the judge. After a long pause Pierre replied, 'I prefer not to answer. I leave that to the experts.'

The green docket, or Certificate of Origin, is, in effect, a French wine's 'birth certificate'. It is stamped with an official seal and issued from a locked machine situated at the vineyard where the grapes have been grown and made into wine. It may be rented out, however, against a deposit, to traders 'of honourable reputation and without previous convictions'.

Pierre solved this problem simply by forming another company headed by his assistant, Serge, another of the accused. Pierre gave

Serge the deposit and told him to borrow the machine and apply the official stamp guaranteeing the authenticity of the wines that they sold. 'There is no fraudulent wine, only fraudulent documents,' claimed Pierre, in a moment of honesty.

Another of the accused, Lucien, admitted lending his trucks to Pierre to transport wine for fraudulent purposes. Serge confirmed that he often saw truck loads of wine from the Midi arriving at the warehouse and that they ended up in relatively expensive bottles of wine labelled 'Bordeaux AOC'.

On other occasions a lorry would leave the warehouse full of an 'inferior' red wine and accompanied by its appropriate certificate. The driver would take the lorry to another warehouse where the certificate would be exchanged for one issued for a similar quantity of white wine that possessed an Appellation d'Origine Contrôlée. The lorry would then return to its original warehouse with the same 'inferior' wine on board but now accompanied by a certificate of authenticity to which it was not entitled. The driver was suitably rewarded for his labours, of course. The AOC certificate did not mention the colour of the wine and since white wine would have in any case fetched a lower price, an increased profit of as much as fifty centimes a bottle could be made in this way, merely by switching the certificates. No wonder Pierre commented on the ineptitude of those who administered the AOC laws. 'How truck ride turned plonk into Bordeaux' was a typical and accurate newspaper headline of the time.

EXPERTS CANNOT TELL CLARET FROM 'PLONK'

This was another extraordinary headline from a national newspaper. Part of the fraud was selling fraudulent wines to the world through a famous firm whose integrity would be thought to be beyond doubt.

Lionel and Yvan passionately and eloquently declared that they were not party to the fraud and claimed that they were the victims of Pierre in the purchase of the wines in question. Both men spent many hours in the witness box and each brilliantly defended their integrity and proclaimed their innocence. They called witnesses to support their allegations that it was very difficult and sometimes impossible to taste a wine and place it accurately.

M. Lionel said that his firm always bought their wine in good faith. He said, 'We always taste our wine with great care, but, for the first time, we are now convinced that we were misled.' He went on, 'It is difficult to taste the wine when it is young. Bordeaux takes time to develop its characteristics.' M. Yvan (his cousin) explained, 'Among wines only six or seven months old and among growths of a different region, confusion was possible.'

Asked how his firm knew what it was buying, he replied, 'I think our house knows what it is doing.' Subsequent revelations by the investigating officers might be thought to give a different meaning to these words!

A Swiss importer of the firm's wines, said, 'Not even chemical analysis can tell where a wine comes from. When the wine is young, it is very difficult. I would not like to be head of a tasting panel.' The Danish distributor of the wines was asked whether he could tell the difference between a wine from St Emilion, close to Bordeaux and an undistinguished wine from the Languedoc in the Mediterranean area. 'That would be very difficult,' he replied.

A director of the British distributors of the wines, told the court that in the seventy years that his company had represented the firm in Britain, they had never received a single complaint about their Bordeaux wines. But the chief investigating officer on the case, challenged this defence. 'In the range of French wines,' he said, 'there are some that resemble each other. However, it was difficult to confuse a wine from the Midi - correct, but without character, without originality - and a Bordeaux wine that, even when young, had a style, an ability to age, that all professional Bordelais were capable of appreciating.'

He was supported by the director of the Laboratoire Départemental de Répression des Fraudes de Tours who said, 'It is impossible for an experienced taster to confuse wines so different.' The defence counsel promptly denounced him as a 'tasting theoretician'!

WRONG ULLAGING

It will be appreciated that wine stored in a cask for any length of time slowly evaporates and creates an air space between the surface of the wine and the interior of the cask around the bung. If this space, called ullage, is not re-filled with wine, the remaining wine will oxidise and deteriorate.

Yvan's claim that his firm knew what it was doing received a blow from a senior inspector of the Service des Fraudes. He told the court that with the ullages practised by the firm in its cellars, no wine had the right to the label 'Appellation d'Origine Contrôlée.'

The inspector, M. Julien le Dorff, giving evidence of his investigations, said that when he went to inspect the cellars he noticed that 'there was only one wine being used for the ullage of all the casks'. He further stated that two members of the family, Alain and Henri-François, told him that they 'used an excellent wine from the Midi' for the ullage.

M. le Dorff continued, 'When one intends to fill up casks of high quality wine in the AOC category properly, one does it from another cask of the same wine, sacrificing the contents of one cask for the sake of the rest.' According to the trial records, some of the casks bore unusual markings, such as 'Bordeaux type', 'Merseault type' and 'can be used for Beaujolais in the United States'.

Yvan responded to these charges by claiming that these wines were used only for the ullage of ordinary wine. He went on, 'Ullage for the AOC wines was always performed by us from the same wines.'

DESTRUCTION OF EVIDENCE

But M. le Dorff would not let up. In a lengthy testimony he alleged that the records that he wished to check had disappeared or were altered during his investigations. Of one analysis book he claimed that it had been 'tampered with overnight while our investigation was going on and with the intention of putting us off the trail. They always produced for us immediately the documents that were completely innocent, but each time we came to something suspicious, then they told us that it had been lost.'

Yvan claimed, however, that the documents mentioned were only for internal use and were destroyed in a normal, routine way when they were no longer required. He also asserted that any changes made in the records were a normal occurrence and no indication of fraud. In a final, impassioned reply to the charges, he said, 'I rub things out and I scratch things out, but I tell the truth.'

He counter-claimed that the complexities of his company's trading were such that the conclusions drawn by the investigators were based on only a partial grasp of the evidence. 'If we had indeed falsified one quarter of our entire stock, as the prosecutor suggests, we would have needed a remarkable number of accomplices among our employees,' concluded Yvan.

M. Lionel said that a new set of stock cards had been prepared and the old ones thrown away to facilitate completion of a qualitative and quantitative inventory!

ALL ON THE BANDWAGON

As the differing frauds and malpractices became evident and the guilt of the accused appeared to be confirmed, the tax division of the French Ministry of Finance entered a demand against the firm for nearly 79 million Fr (£7.2 million). Smaller claims totalling 11 million Fr were made against certain of the other accused.

Then the Trade Associations and Unions entered claims for damages. The National Institute for Appellation Contrôlée (INAO), claimed £18,000 from the firm for falsification of wines and another £18,000 from three of the accused 'for the moral injury that had been caused'.

The INAO lawyer, Maître Boitard, said, 'It will be several years before the prestige of French wine is re-established. We must win back the respect for the quality of French wine.' He urged the judges to ensure that their verdict should '*demonstrate to the world that fraud cannot be tolerated in the wine business*'.

He went on to make it abundantly clear that only a few *négociants* had acted fraudently and that only 14,000 hectolitres of wine out of a total production of about 12.5 million were in dispute. In 1973, he explained, there had been 3,667 checks and none had led to contentious law suits.

The viticulteurs, the actual producers of AOC wines, were next to demonstate their integrity and proclaim their virtue. Their lawyer, Maître Robert Badinter, emphasised the importance of the AOC legislation and the wish to connect the product to the terrain. 'Are not encepage, production per hectare, and methods of vinification watched, to ensure that the product is true and of quality?' he rhetorically enquired. In another place a 'raspberry' might have been heard from the back of the court!

M. Jean Pierre Pomade, counsel for the Union of Southern French Table Wine Producers told the court, 'This case is about the grapes of wrath, the wrath of the Unions when they learn that certain swindlers earned more than four million francs on their backs.'

He continued in the same vein, 'What we cannot accept is that some people who present themselves as the flag-bearers of the profession, who figure in international wine tasting circles as fine connoisseurs, tell us that they were regularly tricked. We don't believe it. We can't accept it.'

UNGENTLEMANLY BEHAVIOUR

Prosecutor Henri Dontenville summed up. 'I want to be the first to say that the wine was not made from artificial ingredients. Absolutely not. But the fraud which consisted of selling ordinary wines with superior labels has been proved and is serious. The responsibility of Lionel is evident; he directed the policy of the firm. The responsibility of Yvan is evident; he was in charge of buying. I do not think one can talk of doctored wines in this firm, but as for the fraud itself, that is to say taking ordinary wines and giving them the Appellation d'Origine Contrôlée, it is not acceptable that such a firm should have deceived or tried to deceive the buyer.'

Maître Dontenville continued fiercely, 'Yes, I am attacking an outfit that can claim a long tradition, which gives it some rights, but which creates some obligations, especially that of being absolutely exact in its commercial transactions; of deserving the credit that the house has.'

His words were greeted with an embarrassed silence.

WHAT ELSE WENT ON

The references to making wine from artificial ingredients and to doctoring wines is interesting at this stage in the proceedings because they had not previously been made by any of the investigating officers. Pierre's first conviction was on two counts. The first was for diluting wine with water. He had been associated with a Jean Dages who physically added the water, albeit at his instigation. Unfortunately, the pump which Dages normally used had run dry and so he used water from a pond that was full of frogs! The second charge, then, was for using water not fit for human consumption. Pierre was sentenced to eight month's imprisonment, but was let off because it was his first conviction.

In the autobiography that Pierre subsequently wrote and inaptly entitled 'In Vino Veritas', he refers to efforts he once made to improve wines by suspending carrots in them. Alas, the system did not work and the wines remained poor.

The actual words 'doctoring wines' may have been derived from the addition of 'vins médecins' which are rich in sugar, or alcohol, or acid, and used to correct a specific deficiency in an ordinary wine. According to Pierre the manufacture of 'vins médicins' can be very lucrative. When sweetening alone was performed, it was necessary also to add some mono-bromo-acetic acid in order to avoid any danger of re-fermentation. Pierre also admitted to having made 'moon wine', again in association with Dages. In the village of Saland, they found a wine vault where they could secretly add sugar to a wine at night, in the 'light of the moon', so that fermentation would continue, make the wine stronger and, therefore, more valuable. One litre of 'moon wine' cost 5.5 Fr compared with 7.5 Fr for a comparably strong wine not so treated – a profit of 2 Fr per litre.

The process proved to be more difficult than it at first seemed, however, because the sugar crystals did not readily dissolve in the tank. Dages had first to dissolve the sugar in water and then add the syrup to the wine in small doses to avoid the wine frothing over and being wasted. Pierre declared that dishonesty was only rewarding through hard work!

Inspector Eugene Gardia, who headed the enquiry, summed Pierre up in this blunt remark to the court, 'Fraud is second nature with him and the entire Bordeaux wine profession knows it.'

The incredible fact is that he remained in business and was never short of suppliers or customers!

THE PUNISHMENT

When the Presiding Judge announced the verdict of the court on 18 December 1974, it was contained in a long and detailed document which he read at speed and then handed to the officials. The result was as follows:

Pierre	1 year of imprisonment with a fine of 27,000 Fr
Serge	6 months' suspended imprisonment, 3 years probation and a fine of 2,000 Fr
M. Lionel M. Yvan François Lucien	1 year's suspended imprisonment, 3 years' probation and a fine of 27,000 Fr
Pierre Servant	6 months' suspended imprisonment, 3 years' probation and a fine of 10,000 Fr
Bertrand Guix de Penos	4 months' suspended imprisonment and a fine of 5,000 Fr

All the others were acquitted. The penal fines were the maximum permitted by law, but in addition there were fiscal fines against the firm amounting to 3,839,000 Fr. Other fines of the same order had also to be paid. Those convicted, immediately lodged an appeal, but on 2 January 1975, the Bordeaux Regional Prosecutor together with the Ministry of Finance Tax Department lodged a counter appeal for stiffer sentences, both against those convicted and those acquitted. The judge remained resolute.

SOUR GRAPES

Informed observers at the trial regarded Pierre's imprisonment as just. He had readily admitted his guilt and his criminal intent. He had organised the fraud on a large scale and involved a number of people in it who might not have fallen had times not been so hard. Much of the presumption of their guilt came from his own fraudulent activities with which they had allowed themselves to become associated. Perhaps that is why so many were acquitted. He was clearly the bad apple in the tub, but many agreed with his remark 'there are thousands as guilty as I'.

He complained of the decision of the dealers' and winegrowers' associations to side with the prosecution so strongly, as 'hypocrisy'. 'I know what they are capable of doing,' he said. Shakespeare might have thought that they protested too much. On the other hand, one might well feel that it was something of an impertinence on the part of the defence counsel to say, 'The wine scandal was an exaggeration for which the Press was essentially responsible and which placed the Bordeaux wine trade, in short, wine itself, in the dock.'

Indeed this whole business leaves sour and cynical thoughts in the mind with scant respect for everyone who had any part in it. It was feared that Appellation d'Origine Contrôlée Bordeaux wine might never again be quite so 'superior' as it has always claimed to be. The Bordeaux Correspondent of the 'Wine and Spirit Trade Record' commented at the time, 'A great wine may cost several times the cost of a good wine for a very slight difference in quality . . . A wine with a famous name may well be inferior to one with a lesser name at a lower price . . . But no one can deny the differences in quality.'

CLEMENCY

There was a delay of only six months before the appeal was heard in July 1975. The findings of the court caused quite a surprise when they were announced, for no one in the trade expected much leniency for the leading culprits after the stern sentences.

Taking into account the damaging publicity suffered by the firm and the fact that an enormous sum of money had still to be paid to the Inland Revenue, the Appeal Court Judges decided to act indulgently. The suspended prison sentences imposed on Lionel and Yvan were quashed. Although they had not actually been to prison, the danger of a committal for some technical breach of the law, ever present in their minds, was removed entirely.

Pierre, the 'Star of the Show', also benefited by the soft-heartedness of the judges. His prison sentence was reduced from one year to six months. Incredibly, he was allowed, even so, to work in his own office during the daytime, supervising and carrying on his normal business. He only returned to prison each night to sleep there until the next morning.

Furthermore, his fine of 27,000 Fr was reduced by nearly three quarters to just 7,000 Fr (around £650). No wonder the trade was surprised. When a self-confessed rogue is treated so leniently for such a major offence, there is clearly little to fear in the adulteration and false labelling racket.

This case was a lesson for many. The Bordeaux region has not only fully recovered but there is a greater demand than ever before; and prices are now higher than ever, too.

23

The Price of Fame

For nearly 300 years, Château Margaux has been producing superb wine. The Médoc classification of 1855 based on the prices fetched for the wine in the previous 100 years, put Margaux in the top four of the Premier Grand Cru.

Château Margaux has been owned by the Ginestet family for the past five generations. They have lived the high life as befits the aristocrats of fine claret, but now times are hard. The present head of the family, Bernard Ginestet, is reputed to have brought much of his trouble on his own head. He has been described as 'L'enfant terrible des Bordelais' since the middle of the 1960s and seems to relish the description. Highly critical of much of the new wine technology and outspoken even beyond his eminent station in the world of wine, he does little to endear others to him.

In the balmy days of the early 1970s when claret prices were rising ridiculously, like rockets, he entered into contracts with some sixty growers in the Médoc and Graves, to buy their grapes and so augment his own range of wines. Although not entitled to the same classification as his own Château, the wines would carry some of the Margaux cachet and so fetch a good price in the market place. Alas, the market soon collapsed. The demand for claret declined both due to the economic situation and to the Bordeaux scandal, but Ginestet still had to honour his contract to the growers and find the cash to buy their grapes, even though he could no longer sell the wine he made from them, at least, not at an adequate price. The depression caused many drinkers of fine clarets to turn to some less expensive growths and quite a number of the owners of the better growths were finding it harder and harder to sell their wine. New outlets were constantly being studied, including mail order and direct selling, e.g. Leoville-Las-Casses, but the going was very tough.

Bernard's next act of folly was in writing a book called *La Bouillie Bordelaise* that Flammarion published in 1975. La Bouillie means 'the mixture' and is a derogatory commentary on the chemical additives used by many vignerons and *négociants* to doctor their wines. He also exploded the hypocrisy, snobbery and vested interests of the hierarchy of the Bordeaux wine trade. The book included poems and verses and was written in a sarcastic style, seemingly endeavouring to be contentious and stir up trouble. Coming on top of the scandal mentioned and all the dirt that that case revealed, the trade closed their ranks against him.

The debts of Château Margaux had reputedly reached astronomical figures, some said even more than £5 million. The Ginestet family seemed most anxious to sell their estate. In 1975, Remy Martin, the equally famous cognac and champagne house made an offer that Bernard promptly refused, declaring incredibly, in public, that 'it would hardly cover our debts'.

The American company, National Distillers, then made an offer, reported to be in excess of £9 million. This was the one that the Ginestet family so badly wanted to accept and asked for Government permission to do so. But the Government declined and refused permission for Ginestet to sell their famous estate to anyone other than a French company.

Château Latour was already owned by a British company, the Pearson Group, which includes the famous publishers, Penguin Books, Longman's, Westminster Press and countless other well known firms. Another Grand Cru Château, Haut-Brion, was also owned by a foreigner, the American Dillon family. Although both the British and the American interests have maintained the prestige of their French estates, the French Government thought that two were enough and firmly declined to allow a third to go abroad. Their national honour - wine - was at stake.

Interest in purchasing Château Margaux's 185 precious acres was slight and offers were few and far between. Another champagne-cognac consortium were vaguely mentioned, but were rebuffed or declined. Then another offer materialised, whether because of some behind the scenes activity of an anxious creditor or from some well meaning Government official, is not known, but it came as something of a surprise and from an unexpected quarter. It was made by the semi-public bank, Crédit Agricôle, whose main business would appear to be making loans to French farmers. Lending a few thousand francs to a small farmer is one thing; buying a prestigious wine estate was quite another and there was much speculation about how the offer came to be made.

'Informed sources' in Paris indicated that the amount offered was 60 million Francs intimating at the same time that it was barely enough to pay the creditors. Naturally enough the family were reluctant to accept an offer so clearly inadequate and pressed in vain for an increase to 80 million francs.

The only alternative seemed to be a public auction of the assets of the estate and there were many hints that this was imminent. The family had reason to believe that they were being sacrificed for the Honour of France.

Then, out of the blue, came another offer – from the Felix Pontin Group, a supermarket chain. Their offer was not disclosed but was thought to be above that of the Crédit Agricôle, although not as much as that from the American Distillers, perhaps £9 million. By then, the financial problems of the family were thought to be so pressing that they had no real option but to accept, and this they did.

Lovers the world over of this fine wine from Château Margaux are well satisfied that the new owners have maintained the wine's high quality and have not, as was feared, sacrificed this great virtue for quantity and cheapness – normally such important factors for supermarkets.

24

Was it Port Wine?

THE ENGLISH DEMAND FOR PORT

It was in 1677, during one of the interminable quarrels with France, that the British Government prohibited the importation of French wines into England. Merchants turned to other countries for their wine and one of these was Portugal. There were already a number of general British traders in Oporto and some of them began to concentrate on wine. Warre & Co, for example were founded in 1670, and Croft in 1678. The red wine available was very full and somewhat harsh, due to the high sugar content of the grape fermenting out. Someone, knowing the sweetness of the British palate then, added some brandy to the wine before fermentation had finished. This had the effect of terminating fermentation and the wine was left with a quantity of residual sugar.

About that time, all Europe vied to buy Constantia, the sweet wine from the Cape of Good Hope. As might be expected, greediness to sell more than was naturally available tempted some of the merchants concerned in that trade to resort to faking and imitation. Its popularity began to fade as the sweet red wine from Portugal became available. In 1703, the so-called Methuen Treaty was agreed providing for the exchange of English woollen cloth for Portuguese wine. Slowly, other now famous firms were established; Offley Forrester in 1729, Sandeman in 1790, Graham in 1814 and Cockburn in 1815. Port wine gradually became more and more popular in England both with the gentlemen and the ladies, at all levels in society. But whenever a wine becomes popular, scandals seem certain to ensue.

The wine is made in the hillsides rising from the valley of the river Douro, a fast-flowing, shallow river, beset with boulders and outcrops of rock.

The vines grow in schistose rock that has been terraced to make a foothold both for the plant and the peasant. At the harvesting, the black grapes are gathered in wicker baskets and carried to the winery on the shoulders of men. Here they are dumped into stone troughs called lagers, where other bare-footed men, trousers rolled to the thighs, trample the berries until each one is broken. Fermentation quickly starts and continues until the wine becomes a deep black/red and about two thirds of the grape sugar has been converted into alcohol. It is then run off into barrels containing grape spirit in the proportion of approximately 9 gallons of wine to 2 gallons of brandy. Fermentation is inhibited by the spirit, the wine retains its residue of sugar and the slow process of maturation begins. In the spring it is taken down the river in small sailing boats to the port of Oporto. Here it is blended, matured in cask or in bottle depending on its quality and then shipped all over the world.

QUALITY CONTROL AND DEVELOPMENT

Back in the hillsides there were, and still are, both good years and bad years. The lifestyle of the peasants was primitive and temptations abounded. It was not unusual to include some black-ripe elderberries to improve the colour of the wine and this continued until the end of the 19th century. Better grape varieties, better care and better methods have now made this unnecessary and it is now illegal to add any other fruit to the grape.

The boatmen, too, had their moments. The journey down river, laden with large casks, was both arduous and hazardous. They helped themselves to wine from the casks and topped them up with water from the river. It was a hard life.

Down in the port, where the wine was blended, further opportunities to succumb to temptation were afforded to the shippers. There are several kinds of port wine, vintage, crusted, tawny, ruby, late bottled vintage, white port and so on. Apart from vintage port which is made from the wine of a particularly good year, the shippers blend together similar wines from different years to produce a fairly standard and recognisably similar wine year in and year out. The older wines give delicacy and the younger wines vitality to produce a delightful balance. Unfortunately a tawny port needs eight years or more, in oak casks to develop its character. This is necessarily an expensive process, since so much capital is tied up for so long. The temptation to blend young ruby port with white port was irresistible to some shippers who thus created a less expensive 'tawny'. A gullible public in England never knew that it wasn't the bargain that it seemed.

The regulations controlling the production of port wine today, were recommended by the Port Wine Shippers Association and enacted by the Portuguese Government, conscious of their need to export high quality wine in exchange for foreign goods. A 'Junta' has been set up to provide the right grape spirit to the wine makers and so ensure that no deleterious spirit is used. No wine may legitimately be called port unless it has been made in a precisely defined area of the Douro valley, and has been shipped out of Oporto with a numbered certificate of authenticity provided by the Government. Anglo-Portuguese Treaties confirm these arrangements and similar Treaties have been made with other countries. France has now outstripped the UK in its consumption of port wine and Germany is a close third. It is undoubtedly a great wine and a fine example will undoubtedly rank high among the premier wines of the world.

THE PORT WINE ASSOCIATION'S CASE

It is against this background that the Port Wine Association brought an action before Alderman Sir Frank Newson-Smith at the Mansion House in the City of London on 3 November 1947. They alleged a contravention of the Merchandise Marks Act of 1887. The defendants were Mr Isaac Bonzer (the manager) and Miss Rose Salond (the proprietor) of a wine business in Arthur Street, London, EC4.

Mr R. E. Seaton, Counsel for the Port Wine Association, stated that on 14 April 1947, a Mr Frank Osmond Morris, a clerk in the employ of the firm of solicitors retained by the Association, bought two bottles of wine labelled, 'Fine Old Vintage Port. Shipped by Feuerheerd Bros. of Oporto'.

Mr F. A. Cockburn, managing director of Messrs Cockburn, Smithes and Co, Shippers of Port Wine, 33, Eastcheap, London EC3, stated 'The wine was not only not port, but also nothing like it!' It smelt of methylated spirits and he described the taste as 'simply filthy'.

The hearing lasted several days and a complicated story was revealed. Mr Bonzer told the court that he was general manager to R. Salmond of Arthur Street. Some time in 1946 he had had a visit from one of his regular suppliers – who offered him some 'old port' at £7 a gallon. He tasted a sample of the wine that a director of the firm had with him and thought that it tasted 'quite nice'. He thereupon purchased three hogsheads (casks with capacities varying between 50 and 60 gallons each). He believed that it was genuine port and had not the slightest reason to suspect otherwise. (It must be remembered that this period was immediately after the Second World War. Nearly all food was rationed and wine was scarce and expensive, since virtually none had been imported for six years. The high price quoted reflects the scarcity value of port wine at that time. Before the war the same wine might have cost less than £1 per gallon.)

In December 1946, stated Mr Bonzer, he was visited by a Mr Kearon of the Layton Bottling Company who wanted to buy some port. Mr Bonzer offered Mr Kearon some of what he had recently bought. After some haggling and since the Christmas sales were over, he sold him 117 gallons at £6.50 per gallon. (If Bonzer really bought the wine at £7 per gallon it is surprising that he sold as much as two thirds of it for £6.50 per gallon, thus losing nearly £60 on the deal, in spite of the time of year!) He strenuously claimed: 'I never at any time described it (the port wine) as having been shipped by Feuerheerd Brothers and in fact had no knowledge as to who shipped it. I never gave anyone authority to so describe it. I bought this wine as port, sold it as port and believed it to be port.'

The original invoice that he made out and gave to Kearon simply described the wine as 'Douro Port'. (This is surprising since he bought the wine as 'old port', a seemingly higher description than the simple 'Douro Port'. Perhaps the wine was originally a very ordinary port that had been lying around a long time.) Kearon claimed that he lost that invoice and while Mr Bonzer and Miss Salond were both out, he obtained from their clerk, a Mr Leaworthy, a duplicate invoice. He suggested to Mr Leaworthy that the wine was shipped by Feuerheerd Brothers of Oporto, and asked for that description to be included on the invoice. Mr Leaworthy thought that because of the price paid for the port, it could well have come from Feuerheerds and included these words on the invoice as requested. He declared that he had been subsequently reprimanded by his employers but that he had acted in good faith.

It next appeared that Kearon had bottled and labelled the wine and sold some of it to a firm trading in Kings Road, Chelsea. It was from this shop that Mr Morris had bought the two bottles of port on 14 April 1947.

One of the bottles had been submitted to Mr W.M. Seaber, an analytical chemist, of Trinity Square, London EC3, on 15 April 1947. Mr Seaber stated that the wine had an alcoholic content of 32.3° Proof (19% by volume) a level that one would expect to find in genuine port. There was chemically no trace of methylated spirit in the wine, although there was an odour a little bit like it.

Mr Edward Hammond, a director of Messrs Feuerheerd, Wearne and Co Ltd, told the court that he had been shown one of the two bottles purchased by Mr Morris and labelled with the name of 'Feuerheerd, Bros. of Oporto'. 'That is not one of our labels,' he declared. Asked by Mr Seaton, the prosecuting counsel, what he thought about the wine, Mr Hammond said that he had tasted the wine and that it was not port nor anything like it. 'Nothing like anything we ship' was his final comment.

Mr Seaton pointed out to the Court that the production of genuine port was governed by an Anglo-Portuguese Treaty. The wine was made from the grapes grown in the Douro Valley in Portugal and fortified with pure grape spirit purchased from the Government who issued a guarantee of authenticity that the wine had come from Oporto. The onus was on the defence to prove that they had no reason to suspect the genuineness of their description of the wine, that they had taken all reasonable precaution against committing an offence and had given the prosecution all the information in their power.

Mr Napley (now Sir David Napley), Solicitor for the defendants, claimed that Mr Kearon should be in the dock instead of his clients, who had all along acted in good faith. They suffered a grave injustice for having to answer the charges made by the Port Wine Association. But Kearon had gone to Jamaica and could not be brought to court.

As his chief witness, Mr Napley called M. André Simon, the President of the Wine and Food Society and author of many books on wine, including port wine. M. Simon told the Court that at their request he had visited the offices of Kingsley, Napley and Co, and one of the bottles in question was produced. It was sealed and had not been opened until he withdrew the cork. He tasted the wine and said that if he had been offered the wine in a bar, he would not have questioned it. He did not think that there was anything wrong with it. He did not claim that the wine was port, but he did say that if the wine were offered to him as port, he would not have questioned it. During cross-examination M. Simon stated that he earned his living writing books and articles. As President of the Food and Wine Society he did not sample or test ports. He had been in Portugal for some weeks in 1920 and had been everywhere in the world where grapes were grown for wine. This honest statement by André Simon, who after all was an expert on champagne and Bordeaux wines, but had not much experience with port wine, started a most difficult time for him. Some shippers of port wine were cross with him and would not speak with him for some time. Worst of all for André, they would not advertise in *Wine and Food* the journal that he published and which gave him his livelihood.

Giving judgement, Alderman Sir Frank Newson-Smith said that the defendants had satisfied him that they bought the wine in the ordinary course of business and that the statutory defence had been established. They had purchased it from a reputable supplier and paid a proper price for it. Their evidence that they sold it in the same condition as they bought it had not been questioned. The summonses were dismissed and the defendants acquitted. Sir Frank declined to grant them costs, however, because he felt that if they had been a little more open with the Port Wine Association, it would not have been necessary for them to bring the case to Court. Mr Bonzer had been too secretive about the names of his customers.

On the face of it, the Port Wine Association would appear to have lost and incurred unnecessary expense on behalf of its members. The wine trade, however, had taken a great interest in the case which had been reported in detail in their press. It was made plain to wholesalers and retailers alike, that the Association was alert to any malpractices. Furthermore, from time to time, sampling was practised and that wherever there was the least doubt, action was taken. Such an attitude clearly protects the consumer and, as far as port wine is concerned, one can be reasonably confident that one is drinking authentic wine from grapes grown in the Douro.

RECENT DOUBTS

Port wine was under the microscope again in 1975. The West German Authorities were investigating some port wine being imported into their country, about which there was some question of accuracy concerning its age. During the analysis, they discovered almost 'accidentally' that the wine had been fortified with a synthetic industrial alcohol, instead of the legally compulsory pure grape spirit.

The producers of port wine, however, are compelled to buy their spirit for fortifying their wine from the Portuguese Government-owned Junta Nacional de Vinho. When pressed by the Port Wine Association for information, the Portuguese Government Trade Office in London, admitted that they had known about this enquiry since the summer in 1974 and that the Government investigations were not completed.

It may be that they never were completed, for the next information that became available was that all of the suspect port had been returned for replacement. There was no question of any danger to health. A spokesman for the Port Wine Association claimed for the suspect wine that 'the quality was just as good'. This could be interpreted as meaning that the suspect port was subsequently sold elsewhere; perhaps in the United Kingdom?

From the legal point of view, the port wine in question did not have the composition prescribed by law. It had, therefore, lost the right to be sold as port wine and was not legally entitled to the name port. Because of the large quantity involved, the Governments concerned did not wish to take part in any action in case they were called upon to foot the bill. There is no doubt that the wine was sold somewhere.

In failing to take public action, the Governments left the way open for further scandals. A future defendant could claim, 'Why pick on me? It has happened before and no action was taken.' When exceptions are made, further abuses are sure to follow and that means more scandals.

25

A Burgundy Scandal

Dijon lies at the northern end of the Côte d'Or in France, and is but a short distance from what many people consider to be the finest stretch of wine country in the whole world. The Côte d'Or (Hill of Gold) is in fact two gently sloping downs. The one nearest to Dijon is the famous Côte de Nuits, home of many of the best known burgundies. The largest town is Nuits-St-Georges, where the 3,000 inhabitants are all engaged in the wine trade. The other down is the Côte de Beaune where all the remaining fine burgundies, white as as well as red, are made. The central town is Beaune where its 18,000 inhabitants work either in the wine trade or for the Church. Many religious orders have for centuries maintained foundations in the area and still provide orphanages, hospitals and homes for the aged or handicapped. Their income is derived from the sale of the produce of their vineyards which they either cultivate or rent out to others.

Dijon, however, is famous perhaps not so much for wine as for mustard and for cassis. A major ingredient in mustard, of course, is vinegar. Unhappily, some wines turn to vinegar if they are not sufficiently cared for, especially from poor vintages. It was to make good use of these failures that the mustard industry was developed. But why cassis? Youngman Carter in his book on Burgundy describes cassis as 'this slightly alcoholic currant syrup'. Julian Jeffs in his *Dictionary of Drink* describes it as 'a blackcurrant syrup commonly drunk with white wine'. Alexis Lichine in his *Wines of France* defines it simply as 'an alcoholic currant syrup' and says that a teaspoonful of cassis added to a glass of chilled white wine is locally called 'rince cochon', which he translates as 'pig rinse' and claims it to be the Burgundian equivalent to Alka Seltzer! Many others enjoy a glass of cassis and cold, dry white wine as an attractive aperitif.

In the wine trade, however, cassis has another, although illegal, use. For poor red wines, or for normally good red wines in a poor year, cassis is a *vin médecin*. It improves the colour, the bouquet, the flavour and the body of the wine. Undrinkable wines become drinkable; unsaleable wines find a market and all thanks to cassis. As the mustard is to the meat, so the cassis is to the wine. But no one does it - unless he's caught!

On the third weekend in November each year, Les Chevaliers des Tastevins de Bourgogne, celebrate 'Les Trois Glorieuses' - three days of tasting the new wines from Meurseault, the Hospice de Beaune and from Vougeot, when they are put up for auction. The prices obtained set the standard for the rest of the wines in Burgundy.

On 23 November 1976, wines from the Hospice de Beaune were auctioned as usual and fetched an average price of £1,188 per tonneau of 900 litres. Deposits were paid and, in due course, delivery was taken by the buyers of the 559 tonneaux of both red and white wines. It is customary to taste the wines again on delivery and when this was done, early in 1977, 107 of the tonneaux were found to contain sour wine - wine which had turned vinegary.

Upon investigation it was found that because of the high cost of new casks (due to inflation they then cost around £60 each) some vignerons had put their new wine into old casks. There is no harm in this provided the casks have been thoroughly cleaned and sterilised before being filled again. Shame to say, they had not. The new wine had been carelessly and negligently put into uncleaned or unsterilised casks. The Hospice had no option but to return the deposits and lost £127,000 of the revenue it uses for its charitable work.

Some of the rejected wine was sent to Dijon for complete conversion to vinegar and for eventual use in the making of mustard. The rest was sent for distillation. The deciding factor was the degree of volatile acidity of each cask. Fortunately, all the scarce and precious white wines were sound.

The Inter Professional Wine Committee of Burgundy subsequently stated that throughout Burgundy, 15-20% of the 1976 red wines had suffered some form of 'unusual alteration' after vinification. The statement was not regarded by many as more than a lame defence of scandalous behaviour.

Countless négociants have their registered office in Beaune or Nuits-St-Georges, usually with a firm of accountants or lawyers. The négociant's address on a label does not mean then, that the wine was grown in that neighbourhood. Similarly, the words 'Mis en bouteille dans nos caves' (or 'nos chais'), means only that the wine was bottled in the négociant's cave or warehouse – not that it was grown in the area.

The total area of the 'vignoble' or grape growing fields on the Côte d'Or is some 9,000 hectares, just under 22,000 acres, but there is always some land that is being rested and the actual area under cultivation at any one time is around 5,000 hectares (12,470 acres). In this small area there are some 15,000 'viticulteurs' producing between 150 and 400 thousand hectolitres of wine each year, of which only half is entitled to the AOC certificate. But these wines are of such superb quality and are in such great demand, that fraud is almost inevitable. Today, most merchants buy Domaine bottled wines for authenticity, or they buy from a reputable shipper. Authentic wines from Burgundy are certain to be expensive. Inexpensive wines from Burgundy are unlikely to be authentic.

In reviewing Anthony Hanson's book *Burgundy*, Christopher Fielding quotes, '... too much praise has gone to the Burgundians' heads. The fabled quality has almost ceased to exist, even at outrageous prices.'

Mr Fielding goes on: 'No great name of Burgundy is free from attack.'

There is of course much truth in what Anthony Hanson says. The growers and merchants of Burgundy have for long had an easy life. Their wines have been a commodity in demand; they have been able to ask for high prices ... and get them. In Britain, particularly, the late arrival of the application of Appellation Contrôlée laws has meant an easy outlet for any wine with any name. Indeed, even now, there are several members of the British wine trade who believe that the introduction of such legislation is the worst thing that has happened for centuries.

The French authorities, too, are to blame. The inspection system is totally inadequate and too many prosecutions are made or not made, for purely political reasons. (I once had the misfortune to be harangued by a socialist député for suggesting in a speech that over-chaptalization spoiled the charm of Beaujolais.)

26

The Wine That Never Grows Old

The name Beaujolais is, perhaps, as well known and popular as Sauternes. For those who are never sure what wine to drink with the meat course of a meal, red Beaujolais wine is a favourite standby. At one time it was very inexpensive and could be found on the wine list of every licensed restaurant, both in this country and many others. Its popularity in France was such that it has been said that the Parisiennes alone drank all the wine that Beaujolais produced. It has also been said that more Beaujolais wine was imported into England than was produced in Burgundy. A thousand years ago the Emperor Charlemagne forbade the storing of any wines in Burgundy that were not produced in the district. The implication was evident that the Burgundians were tempted, then as now, to 'stretch' their wines.

A similar argument applies to the variety of vine grown in the Département. The noble vine for the red wine of Burgundy is the Pinot Noir, a somewhat shy bearing variety, albeit producing wine of superb quality. The temptation is to dig up the Pinot Noir and replace it with the Gamay, a prolific cropper, although it produces a less aristocratic wine in large quantities. In the Beaujolais area of Burgundy, however, the Gamay does better than elsewhere and its light, fruity wine is the reason for its popularity.

Throughout the centuries and in many wine districts, festivals are held each year as soon as the busy work of pressing, fermenting and racking the young wine has been completed. Indeed, the festival is all about the new wine which is then drunk in copious libations. Beaujolais is one of those wines that often tastes best when it is very young and fresh, full of the fruity taste of fat grapes.

In the middle 1960s, some of the bright young men and women of modern society in London decided to race each other home from the Beaujolais, carrying as many cases of the new wine as they could manage. The *Sunday Times* offered a trophy for the winner of the race and from there it grew. The competitors are not allowed to leave the Beaujolais area until midnight on the third Thursday of November, the earliest date on which the new Beaujolais wine is allowed to be distributed. The object of the race is to get the Beaujolais Nouveau into the London wine bars as early as possible. The finishing point is the Houses of Parliament and prizes are offered for those flying private aircraft, for those driving large cars, for those driving small cars and for those driving motor cycles, although the quantity that can be carried on this vehicle can hardly be more than a single case of a dozen bottles. What the wine tastes like after such a buffeting and without a rest is a matter of opinion. The race is fine for a bit of fun, like the Shrove Tuesday Pancake race or a carnival wheelbarrow race; all harmless fun and enjoyable for participant and spectator alike, but it does nothing for the dignity of wine. There are dangers, too, in racing on British roads and in 1977 the Police Authorities banned the race as such. This did not deter certain individuals who scurried back with the very young wine so that they could recount how they had drunk the wine on the very first permissible day. There would seem to be no other explanation. Here is a wholesale offer:

'We guarantee (subject to an Act of God) delivery on Thursday 21 November 1985 if you are located either 150 miles from the centre of London or 85 miles from the centre of Leeds. All other UK locations will be delivered on Friday 22 November. Collections by arrangement.'

But what of this Beaujolais Nouveau or, more precisely, Beaujolais *en primeur?* In a very good year (such as 1978) the wine can be a delight to drink, but in most other years the quality is such as to question the growers and *négociants* themselves for producing such an unpalatable beverage. But once again, the temptations to do so are enormous. Most wine growers have to wait many months at least and sometimes years, before their wine has matured enough to be able to sell it. With Beaujolais Nouveau, the money starts to come in after the November following the harvest, long before that reaching growers elsewhere.

The minimum strength of Beaujolais is 9% alcohol by volume and for Beaujolais Villages the maximum strength is 13% – as a rule the actual alcohol strength is 12.5% with an acidity level of approximately 0.5%. Chaptalisation by 2% alcohol by volume is allowed. Wines, which after the permitted chaptalisation have an alcohol content still under 9% lose the appellation and become *vins de table*. Actually the vineyard must be at least four years old for the appellation Beaujolais.

There is a special technique in producing this wine that has benefited more than most from modern technology. Most important in the treatment of Beaujolais destined for sale as Nouveau is the use of especially good and quickly working yeast. Only a few of the grapes are crushed and yeasted, the rest are piled, unbroken, on top and the vat is sealed. The carbon dioxode released from the fermentation of the crushed grapes causes a change in the cell structure of the uncrushed grapes and some intra-cellular fermentation also occurs. As a result the colour in the skins of the uncrushed grapes is released into the otherwise colourless juice. This process is called *maçeration carbonique*, and is used in some places in the south of France, in California and to some extent in South Africa and Australia as well. Although the colour is readily extracted by this method, less tannin is extracted and so the wine takes less time to mature.

The must is very lightly sulphited (5 to 8 g per hundred litres) and warmed to between 23° and 24°C, and the maceration is continued for five or six days. By this time the fermentation of the free-run juice is now nearly completed and that in the uncrushed grapes is about half finished. The entire content of the vat is then strained, the uncrushed grapes are crushed and pressed, the juices are combined and fermentation is continued and finished within fifteen days of harvest.

Next, the young wine is racked from the sediment and a malo-lactic fermentation begins almost at once, converting the sharp malic acid from the grapes into the milder tasting, lactic acid. Within a few days the wine is centrifuged and filtered to clarify it and remove the yeast cells and other micro-organisms. Next, it is treated with metatartaric acid, to prevent the crystalisation of potassium bitartrate in the bottle, and the wine is degassed to regulate the quantity of carbon dioxide. Too much CO_2 causes the wine to taste prickly; too little and it tastes flat. The right amount contributes to that fresh, fruity flavour.

The wine is again filtered, gum arabic is added to prevent flocculation of the colouring matter and the wine is finally filtered and bottled. With any luck all this is completed before the third Thursday in November. In very bad years, such as 1977, the task is impossible, although in 1984 it was as early as 14 November. As a result the release date is put back a week or even two, so that the vintner has a few days longer to complete the wine. From time to time, certain growers think more of their integrity than that of the waiting cash and decide not to produce an *en primeur* rather than produce a bad one.

Although a great deal of fuss is made about Beaujolais Nouveau in the English and German press, the largest share of this wine stays in France and many Frenchmen like to know that the Beaujolais Nouveau has arrived. A substantial share goes to West Germany and another to Belgium. The United Kingdom ranks fourth among the devotees of this wine.

According to a report of the Union Interprofessionelle des Vins du Beaujolais, the export of Beaujolais Nouveau in the months of November and December from 1972–84 was as follows:

Year	*Hectolitres*	
1972	71,328	(out of approx. 360,000 hl. of Beaujolais for the whole year.)
1973	60,626	
1974	93,579	
1975	90,982	
1976	77,468	
1977	68,726	
1978	98,719	
1979	124,292	
1980	152,850	
1981	154,254	
1982	158,951	
1983	217,513	
1984	226,389	(out of approx. 565,000 hl. of Beaujolais for the whole year).

About 50% of all the Beaujolais wine now produced is made into *en primeur* and should be drunk while it is still very young. This chapter is called 'The Wine That Never Grows Old' because by the following vintage, the wine is no more *nouveau* but as dead as yesterday's newspaper. The treatment that the wine receives in its making is such that it does not live longer than one year. All the traditional methods of producing quality wine are transgressed for the sake of an arbitrary date.

The merits or otherwise of making an *en primeur* are discussed year after year in the trade and national press of the United Kingdom. It would seem that the main, if not the only, merit of this wine is that of fashionable snobbery. To be seen drinking Beaujolais in mid-November carries cachet in modern bar society, whether in this country or on the Continent.

There are a number of imitators now and the Swiss are advertising their Beauvalais Nouveau. Further south, growers on the Côte du Rhone and in the Corbières district, and growers in Italy too, are also thinking of how they can capture some of that early money. To hell with quality! The latest candidate for 'Nouveau' is South Africa. Since they harvest their grapes and make their wine in what is the European spring, they can offer their vintage long before the Europeans.

To those who feel they must take part in this game I give the following advice: Open the bottle and drink one glass daily. You will find that the wine tastes much better on the 3rd, 4th or 5th day. Airing improves this wine, so drink your next bottle according to your findings and increase your enjoyment. Frankly speaking, I do not enjoy grouse shot on 12 August for my dinner on the same day, and I do not drink Beaujolais Nouveau when it is really nouveau!

27

The Ipswich Fountain

Boastful of their reputation for stocking the widest variety of wines in the world, English wine merchants have for the most part abided by the Merchandise Acts.

Some two to three centuries ago, home-produced fruit wines were added to wines knocked about by a rough crossing from Bordeaux. The revivified wines were subsequently sold as 'superior quality'. Not having indigenous grape wines with which to compare the imported wines, the rich merchants and aristocracy, who alone drank wine, grew accustomed to flavours different from the original wines described on the labels.

The frequent wars with France caused the supply of wine to be intermittent and merchants often had to take what they could get rather than what they wanted. With the development of the Industrial Revolution and the growth of the middle class, wine snobbery became a dominant consideration in the marketing of wine.

As long as the label described the wine as sufficiently appropriate to impress one's guests the precise contents of the bottle were not of significance to most wine buyers.

Many wine merchants, especially of the 'here today, gone tomorrow' type, labelled their bottles accordingly, St Emilion, St Julien, Nuits-St Georges, Beaujolais, Châteauneuf du Pape and so on. The contents were often a blend of inexpensive red wines. Later, when the Appellation d'Origine Contrôlée regulations were instituted in France and became the hallmark of authenticity, the labels were printed with the name of the appropriate AOC designation. Who could prove otherwise? Nicholas Tomalin, a journalist with the *Sunday Times*, could and did.

In the autumn of 1966, he heard about a firm called the Société des Vins de France whose English headquarters was a warehouse-type building on a quay at Ipswich harbour. The main contents of the building were some large tanks, heavy canvas hoses and pumps. On one side, however, there was a line of bottling, corking and labelling machinery.

Several times a week a small coaster would tie up alongside the quay and unload a number of 550 gallon bulk containers. The wine would be pumped into the tanks and, in due course, the containers returned to the coaster.

Originally, the set up had been an orthodox extension of the business of the parent company. In France the Société des Vins was one of the largest companies supplying *vin ordinaire*, under the brand name Valpierre, to French families. Under French law they could not use specific or even regional names for their wines, hence the brand name. In fact, they acknowledged that the wine came from Algeria, Provence and the area around Marseilles, areas whose wines were better known for their cheapness rather than their quality. Nevertheless, the skill of the management in blending and marketing the wine was highly successful and the company prospered and extended their operation to the UK.

Now French families accept their wine in a litre bottle with an easily removable form of a crown cap. After all, to them it is only their daily wine, not the nectar of the gods. Valpierre was a sound, drinkable wine, and jolly cheap too! What more could one want for everyday drinking?

In Great Britain, however, wine is regarded differently. Although it is now bought and drunk by a wider range of people than before, it is still regarded as a luxury, usually to be drunk only on Sundays, high days and holidays. People naturally want their wine to be corked and labelled with a familiar and orthodox name. They lack the confidence of their palate and want to be reassured by the label that they are drinking wine of good repute and quality. Valpierre was a meaningless name and the wine might have come from anywhere - and as we now know, it did!

The French management were disappointed with their lack of success in marketing their popular Valpierre in Britain and, at first, failed to understand the snobbery of the wine customer here. Determined to succeed as an enterprise, however, they began to supply their wine in bulk quantities to those merchants who were prepared to accept the wine simply on its taste value. No questions were asked on either side. Not all the interested merchants had bottling facilities and some asked for the wine to be bottled and labelled for them with their own labels at the Ipswich depot.

Both kinds of merchants were supplying to their customers

wines that had no relation to their label. Some cognoscenti had suspected this for many years but previously most wines were bottled and labelled in France and it was virtually impossible to obtain any tangible proof to the contrary. The British wholesaler would merely shrug his shoulders and declare that he bought the wine in good faith and trusted his French supplier.

Thanks to the success of Valpierre in France, there was now a blending organisation operating in the UK. It might now be possible to enquire into the ethics of the wine trade, or at least some of them. Nicholas Tomalin declared, 'It was only necessary to devote some time to careful observation and questioning to discover precisely what was going on.'

Mr Tomalin and a few colleagues observed and questioned tactfully and inoffensively. They discovered and reported in incontrovertible detail an almost unbelievable story in *The Sunday Times* on 27 November 1966.

The manager of the Ipswich depot of the Société des Vins de France was a Mauritius-born young Frenchman called Michel de Fontenay. His head cellarman and accounts clerk were English and so were his four or five labourers who handled the containers, hoses, etc. Eight basic wines were being imported, two reds, three whites and three rosés. Some of these were sold in bulk, and Michel de Fontenay himself was somewhat shocked to find in a restaurant a rosé costing 27/6d a bottle which he had supplied to the proprietor at 10/- a gallon - enough to fill six bottles!

He allowed the reporters to follow the two reds, (they were called the Eleven Five and the V.P.) through the often complex blending operations. Eleven Five simply meant 11.5% and referred to the alcohol content of the wine. It was, he said, a mixture of wines produced in the Languedoc area of the Midi in France, where most of the French *vin ordinaire* is produced.

This wine sold successfully in France under the name Kiravi Export and at that time cost only 1.80 Fr per litre - not bad for a pleasant but unpretentious wine.

V.P. meant Vieux Papes and had nothing to do with the well known British Company where the V.P. meant Vine Products and fooled no one. Vieux Papes had an alcohol content of 12% and was defined by the Société as 'an unclassified wine of superior quality'. It was claimed that it was a blend of two classified regional wines, but these were never specified. It should be noted that in France the adjective 'supérieur' is customarily applied to wine containing 12% alcohol, e.g. Bordeaux Supérieur. Nevertheless, Vieux Papes sold in France at more than 7 Fr a litre – quite a high price for a branded wine in 1966 or even today.

When the Ipswich organisation was set up at a cost of £30,000, a subsidiary marketing organisation was established and called Société Vinicoles Française (UK). The aim as before was to market the brand wine Valpierre. When it was seen that the endeavour would not be successful the subsidiary was sold to a gentleman in Hertfordshire, who was marketing wine on a mail order basis. Not unnaturally, he was a major client of the Société des Vins de France but by no means its only customer. Many well known names were listed in their order books.

Under the heading '*Selected French Wines – Red*' the mail order list offered the following attractions to its customers:

Beaujolais, 100/– per dozen, 8/4d per bottle. A light smooth wine from Southern Burgundy. Should be drunk young, with almost any food. Can be served cool.

Nuits-St Georges, 126/– per dozen, 10/6 per bottle. A fine wine from the Côte d'Or of Burgundy. Quite full in body, well balanced and excellent with any kind of meat.

Châteauneuf du Pape, 132/– per dozen, 11/– per bottle from the sun-baked valley of the Rhône, just to the north of Avignon. A big, full wine, it goes well with any meat, especially game.'

Nicholas Tomalin was able to see how these wines were produced. The first order was for 50 dozen bottles of Châteauneuf du Pape. M. de Fontenay merely arranged for the tank attached to the bottling plant to be filled with V.P. The wine was bottled, corked and labelled (from the supply already delivered), cased and set aside for despatch.

Next on the list was the Nuits-St Georges. The labels in the machine were changed and then a measured amount of the Eleven Five was pumped in and so mixed with the V.P. already in the tank. The measure was a humble dip stick. It got the quantities about right, since a little more or less would not be noticed. In any case the wine had now been transformed into 'Nuits-St Georges'! The machine was switched on and automation did the rest.

Finally an order for Beaujolais was fulfilled. After removing the 'Nuits-St Georges' fancy label, which had a seal implying authenticity, an even fancier label with a shield emblazoned as a coat of arms was placed in position. Some more Eleven Five was pumped into the tank so that the content was now roughly two parts Eleven Five to one part V.P. Once more the machinery whirled and the bottles jangled to produce what gullible snobs would innocently believe to be Beaujolais. Ironically the 'Beaujolais' contained the least amount of the Vieux Papes which de Fontenay sometimes called his 'Beaujolais type'. The Eleven Five he irreverently referred to as his 'Médoc', or 'claret type'. How fooled can you be!

When subsequently asked about his French reds, a spokesman for the mail order firm stated, 'I am very happy to offer these three wines as very good value and typical of their areas. I know where they're from.' In the context of his remarks the naïve might assume that he was referring to France, but perhaps he meant what he said.

But what *is* a typical Beaujolais, Nuits-St Georges or Châteaunuef du Pape? Well, it depends on your palate. M. de Fontenay's customers differed in their tastes. Some would want more Eleven Five, others more V.P. in their blends. Each had their own idea and de Fontenay did his best to please everyone and made up blends to meet their individual idiosyncrasies. He commented to Mr Tomalin, 'You English are such individualists. Each one wants a different blend for a different wine. No one has the same palate ... Very confusing. As a result I have to keep halting my bottling line, pumping wines from tank to tank, wasting gallons – just to satisfy old-fashioned quirks. It was not what this operation was designed to do.'

He was right. The operation was designed to sell to people in Britain a sound and pleasant wine under the brand name of Valpierre. His potential customers, however, thought that Valpierre sounded too common and preferred to pay rather more for the same wine with better class labels. It made them believe that they were drinking something better.

The scandal here is not so much the activities of the various wine merchants, but rather our own snobbery. In the year 1966 the consumption of wines in the United Kingdom was only four bottles per head of the whole population. In the year 1976 the consumption had increased to only nine bottles per head. The French consumption was then static around 106 litres per head, say 140 bottles per head of population. But then the Frenchman pays hardly any tax on his bottle of wine and nothing comparable with the current United Kingdom minimum of 60 pence per bottle. If they had to pay five francs tax on every bottle of vin ordinaire, the French consumption of wine might decline dramatically, or there would be a national revolt instead! Marie Antoinette is reputed to have said when the masses revolted before, 'If there is no bread, give them cake instead.' For more than 2,000 years French people have been accustomed to wine. It is unlikely that many of them would ever give it up for water - even from a spa - although some have. Wine is as natural and as precious to the French as their splendid bread.

Wine snobbery is now much less than it was. The improvement in the quality and the ready availability in the supermarkets of the less expensive table wines enables many families, especially in the more prosperous south of England, to enjoy wine with their meals with some frequency if not regularity. The increased popularity of dining out is almost always accompanied by the drinking of wine. At social functions, too, wine is quite commonly served.

Post-war United Kingdom governments still believe, however, that only the better off drink wine, and continue to regard wine as eminently suitable for double taxation - Excise Duty and VAT. M. de Fontenay and many others were guilty only of indulging the whims of a few people in what is in Britain still something of a luxury. Had they not done so, some people for sure would not have been able to afford to drink wine.

28

The Italian 'Phoney Wine' Scandal

Wine has been made in Italy for perhaps 3,000 years. It is thought that the Phoenecians first planted the vine at their coastal landfalls and that over the centuries the cultivation spread inwards. In any event, the Greeks referred to the land that we know as Italy, as 'Enotria' or 'Land of Wine'. By the time the Romans were in widespread occupation, wine making was well developed and wines such as Falernian, Sorrentan and Setia were popular and highly appreciated. Indeed, to prevent profiteering by the wine merchants, the Emperor Diocletian in AD 301, fixed a price differential between the famous wines and the everyday wines. The best wines could be sold for 30 deniers and the everyday wines for 24 deniers. He even threatened the wine growers and merchants with nationalisation if they did not keep their trade in order.

Italy is now the largest producer of wines in the world. Her annual output is between 75 and 78 million hectolitres (1,650 and 1,716 million gallons). Her population consumes 55 to 58 million hectolitres a year, or 85 litres for each man, woman and child. Of the remaining 20 million hectolitres 5.8 million hectolitres goes to France, 4 million hectolitres to Germany and 1.1 million to the UK, and the remainder to the rest of the world. Some of the wine exported to France and Germany is used for blending with the indigenous wines to give them some body and strength. The increasing quantity imported into the United Kingdom is mostly served as carafe wines in Italian restaurants or sold in jumbo-size bottles in supermarkets and cut-price wine shops. In general terms, Italian wines have not yet earned a high reputation among British connoisseurs, although it is readily admitted that the standard of Italian wines is rising steadily and that the wines are inexpensive.

This is the background to a scandal involving many hundreds of people engaged in the large-scale manufacture, marketing and distribution of fake wine. Furthermore, even though the people concerned knew that the wine had never been near a grape, let alone contained any grape juice, they advertised it on TV as 'genuine wine'.

On 10 November 1966 *The Times* carried a short paragraph mentioning the arrest of some thirty people in different parts of Italy whom, it was alleged, had been involved in passing off 'phoney wine' - a phrase the journalists subsequently used quite often to describe the liquid produced and sold as wine.

The arrests had been made in a series of raids in towns as far apart as Brindisi in the very heel of Italy, and Reggio Emilia in the distant north. The phoney wines, labelled with the names of Italy's most famous wines (such as Barolo, Lambrusco, Chianti, and Frascati) had been produced in highly mechanised plants in places such as Frosinone, Arpino, Naples, Ravenna, Fressico and, of course, Brindisi and Reggio. 450 tons of the phoney wine had been confiscated at Frosinone and Brindisi alone, together with 120 tons of artificial flavouring and an illegal sugar solution. More raids were expected together with further arrests and confiscations.

Many months later in 1967, the hearing began in a tiny court house in Ascoli Piceno, a coastal town on the Adriatic. Why this inadequate venue was chosen is perhaps another scandal, for the charge was serious enough: 'making, marketing and distributing fake wine'. The person primarily concerned was said to be one of Italy's best known wine manufacturers. The number of accomplices in the dock had grown to a staggering 280. When you also consider that the wine trade in Italy employs 7½% of the country's work force, it is surprising that the case did not receive greater publicity in the foreign press. Remarkably little of the case was reported in the British national or wine trade papers, although the hearing dragged on for more than a year. Perhaps this was a reflection on the reputation of Italian wine some years ago. Maybe no one was interested - not even the wine trade.

During the dreary hearing in the hot and stuffy court room, certain revelations were made, albeit with great reluctance. It appeared that millions of gallons of phoney wine had been made and sold, particularly in the supermarkets. The phoney wine was said to have been made from 'a mixture of tap water, sugar, ox blood, chemicals and the sludge from banana boats'. Somehow or other the defendants had devised a way of making their concoction in as short a time as 8 hours! This was moonshine without a doubt.

When the anti-adulteration police squad arrested the first of the gang, some 770,000 gallons of the phoney wine were also taken into custody and stored in vast vats in a 'secure' place. From the size of this stock it was clear that a great number of people must have been involved. Tracing so many and proving each individual connection would necessarily take a long time, but there seems to have been little sense of urgency, either on the part of the police or the government authorities, to bring all the rogues to justice.

During his examination in court, the well known manufacturer of that 'genuine wine' that he advertised on TV declared: 'I may be the king of the wine fakers, but there are many others besides me and not just in this court room. Everyone knows that.' Similar sentiments were to be uttered by the organisers of the Bordeaux scandal a few years later.

A government official, Signor Alto Dolciumi, commented: 'Wine faking is a flourishing industry in the Romagna province. At night they dump sugar, rough wine and coloured water into underground tanks. A few days later, after the miracle of fermentation, trucks appear and take the mixture away. Sometimes they use fish gelatine, powdered blood and other additives to enhance the taste. The fake wine is very competitive.' Indeed it was, at the equivalent of 11 pence per litre!

After so much and so long a non-event, with 'evidence' from as many as 1,200 witnesses, the trial died a sudden death. The judges unexpectedly terminated the proceedings and ordered a re-trial. The explanation given to the public was to enable the police to widen their investigations to include tax evasion charges against the accused and associated accomplices not yet arrested; further, to make such a large scale trial as would put an end for ever to the faking of wine! Those 'in the know' anticipated that the number eventually accused might even reach 1,000 persons. The new trial was scheduled to begin in the spring of 1970, but, of course, it never did.

Bold words were uttered by the legitimate wine industry who felt that they were suffering for the sins of the guilty. The Government was urged 'to strike without pity against the fakers'. But the Government had other matters on its mind – like politics

- and was incredibly loath to do anything like governing. One cannot help but wonder who was getting at whom.

The court officials told those who enquired when the re-trial was going to begin, 'Investigations are still going on in a particularly laborious investigation. So we have no idea when the new trial will take place.' The anti-adulteration police squad spokesman said in self defence: 'We can't be everywhere all the time.' But then, everyone knew that the unit was far too small and ill-equipped to combat such a widespread and well organised racket. Indeed, they could not even guard their most precious exhibit. Perhaps that is the real reason why the court never reached a conclusion, nor ever passed judgement on the self-declared 'King of the Wine Fakers' and his countless accomplices. For someone, at some time unknown to the authorities and without leaving any clues behind him, stole the 770,000 gallons of fake wine! Quite a large fleet of tankers with a number of men must have been used to move so large a stock of wine. It weighed around 3,125 tons. It would have taken much time, too, to fill and empty the tankers. Someone with great authority and power over others must have been in control, for no one saw anything, heard anything or knew anything about it. Silence was total.

Only one lonely voice was raised on behalf of the duped consumers of the phoney wine. Signor Vincenza Dona, the President of the National Consumers' Union, declared, with naïve euphemism, 'Whether all the wine we drink is real wine is an open question.' Signor Dona attempted to lead a movement to protect the 'good name' of the Italian wine industry. He wanted to have published a detailed catalogue of all the authentic wine labels used by the legitimate trade. He urged, too, that the wine trade as a whole, growers, merchants and retailers alike, should police themselves and root out the unscrupulous element.

Money was provided to produce the catalogue and organise a secretariat. But producers and merchants are highly individualistic. They may describe their wines by the name of the main grape that they use, or by the name of a nearby town, or both! They may use a fancy name that they have made up, or a legendary name, or one with an historical reference or even a brand name. Sometimes the same name is used to refer to several different wines, red or white, sweet or dry. They could hardly be less interested in Signor Dona's ideas. A bureaucracy that puts off until tomorrow anything that can be done today, hardly exudes encouragement to growers and merchants to put their house in order. In the meantime, the money remains in the Government Treasury.

29

False Vermouth

Very ordinary wine, often not readily saleable, is used for making Vermouth. It is frequently a blend of otherwise dull wines, flavoured with a mixture of herbs such as cloves, camomile, hyssop, quinine, juniper, coriander, orange peel, and, of course, wormwood, from which it gets its name (vermut in German). Spirit is added to increase the alcohol content of the wine to around 17% – often described as 30° Proof. The precise blend of herbs varies from one manufacturer to another and is kept as secret as possible.

Martini Rossi is one of the best known manufacturers of this very popular apéritif. They not only market their own brands under their own name, but also supply their wine in bulk to certain customers who bottle and label the wine themselves. Being a very considerate firm, Martini Rossi supply the labels already printed and in very generous quantities in case the customer wishes to use some half size bottles or spoils some in the labelling process.

One can imagine their concern and hurt when they came across a bottle labelled Martini Rossi Vermouth in the orthodox manner, but containing a British-made Vermouth, not at all like theirs. They determined to find out who had misused their services.

Tracing the culprit proved a difficult and expensive task. New labels had to be printed for every customer, each with a separate and secret identifying mark. The labels had to be supplied in the ordinary course of business as required by their customers and without alerting them in any way. Then the bottles of wine from each customer had to be monitored and analysed until the scoundrel was found. He turned out to be a grocer with a particularly large turnover in Vermouths. He had obtained a very cheap supply of an inferior Vermouth from an unnamed source, bottled it, labelled it with the spare Martini Rossi labels to save printing his own and sold it for the same price as the real Martini Rossi Vermouth.

Naturally, Martini Rossi stopped his bulk supply forthwith and thereafter fulfilled his orders with case lots of ready bottled and labelled Vermouth. The grocer's customers who had for so long been cheated and duped into thinking they were drinking something which they were not, were again able to drink exactly what they purchased. The grocer was very fortunate in not being prosecuted and punished. In another case the culprit was not so fortunate. He was forced to leave the wine trade and all the riches he had assembled were lost.

Martini Rossi went even one step further to clear their name and restore their reputation. No Martini Vermouth is now supplied in bulk to anyone. There is only one bottler: Martini Rossi themselves.

30

Prohibition in the USA

In the minds of many American people, Prohibition in the USA was the greatest scandal of all time. It was imposed on Americans in 1919 and lasted until 1933. The Federal Law prohibiting the manufacture and sale of all alcoholic beverages throughout the country gave a very special meaning to the word 'prohibition', so that it now immediately brings to mind that sad state of American affairs at that time.

The Great War of 1914–18, the war to end all wars, had cost such an enormous toll of human life, that upon the termination of hostilities, the Allies were full of a moral urgency 'to make the world a better place to live in' as though this would in some way justify the four year holocaust. In the United Kingdom it was expressed in the term 'A Land Fit for Heroes' and such a land included at least beer and spirits for the workers and a modest amount of claret and port wine for the Colonel Blimps in the London Clubs.

In the USA, however, the Methodist Ministers from the 'Bible Belt' in the Deep South, led a tremendous campaign lobbying Congressmen and Senators to abolish the American saloon. Even many people less opposed to the drinking of alcohol than the Methodists regarded the saloons as 'sinks of iniquity'. To many Americans, the million soldiers coming home from the war were the saviours of mankind and had to be welcomed into a clean-living, sober and industrious society, that put its emphasis on an upright attitude to work and to moral leadership. It is doubtful whether the soldiers agreed on either point.

The saloons, alas, were often the only amusement in the neighbourhood and attracted ne'er-do-wells, loose-livers, prostitutes, thieves and alcoholics, collectively called 'the dregs of society', as well as those ordinary, normal people just in search of a drink to slake their thirst. Recognising the evils, both real and potential, of the saloons, of 'Demon Rum' and all the other alcoholic beverages, the euphoric Federal legislators bowed to the 'do-gooders' and passed, with but little discussion, the necessary amendment to the Constitution prohibiting the manufacture and sale of alcoholic beverages. The decision was soon ratified by more than the necessary two-thirds of the then forty-eight States and America became 'Dry'.

But not everyone was in favour of such drastic action. Cardinal Gibbons of Baltimore, a much respected leader of the Catholic Church, denounced the decision as 'inhuman'. Some City Authorities declined to enforce this particular law that was so unpopular with many ordinary folk.

Almost overnight, reaction set in. Entrepreneurs quickly brought over shiploads of whisky from Scotland, rum from Jamaica and wine from France. The ships anchored three miles offshore in international waters and by night the 'lush', as it was called, was unloaded into small vessels that could smuggle the drink into quiet coves and from there into waiting trucks, ready to drive straight to the Speakeasies. The Speakeasies were, effectively, private drinking clubs of which you had to be a member, or known to a member, to be admitted. The swing doors of the saloon had been replaced by stout oak doors with locks that could temporarily hold back the Federal Enforcement Officers. The name Speakeasy was developed from the Irish expression 'Speak easy', words used by the barmen to their customers. In other and more verbose words 'Talk quietly so that your voice cannot be heard by a passing patrolman'.

Sometimes the owner of a Speakeasy would brew his own beer and even distil his own hooch. Although the profits were greater, the risk was greater still and the vast majority of owners bought in their illegal stocks from the black market. All the buying and selling was done for spot cash. No records that might incriminate them were, therefore, kept and consequently no tax was paid. Gangs soon organised the distribution of the booze. Each gang was headed by a Mafia-linked Italian overlord such as Frank Costello, Lucky Luciano and Joe Adonis who controlled the eastern coastal States. In the Midwest, the infamous Al Capone reigned supreme, albeit associated with the Mafia. Rival gangs of Irish and German criminals were led by Legs Diamond, Dutch Schultz, Mad-Dog Coll, Boo-Boo Huff and Oweny Madden. Often the rival gangs would hi-jack consignments or pirate the ships, shooting it out with each other in typical gangster fashion, as subsequently depicted in the Hollywood films. The police left them alone, partly because they were too afraid to get involved, and partly because they thought that it was simpler, albeit rough justice, when the gangsters killed each other. Quite frequently, however, the police were in on the racket and even tipped off the proprietors of the Speakeasies when the Federal Officers were planning a raid.

Printers carefully duplicated the labels of well known and reputable brands of liquor and supplied them to the makers of the illicit booze. Even so, the critical palates of many drinkers were able to tell the difference between the illicit and 'the real McCoy'.

Bill McCoy was an ex-Officer of the United States Merchant Marine who had trained under the venerable Admiral Sims. He bought a small fishing schooner and carried 1,500 cases of spirit from Nassau to Georgia for a fee of $10 per case; a highly profitable first trip. Soon he bought a larger schooner and started regular runs from the West Indies right into New York City and other large coastal towns. He only carried properly distilled and genuine spirits from orthodox suppliers, hence his goods were always authentic and never false or of poor quality. As a result, the genuine goods became known as 'The real McCoy'; a term that was subsequently applied to any genuine article, and has remained in our language ever since.

After a while, the people frequenting the Speakeasies began to get choosy and demanded better quality drinks and surroundings. They refused to pay the huge sums demanded by the gangsters. The cost of drinks in the USA fifty years later was less than that charged during the prohibition. By the end of the 1920s the position had improved considerably. Chairs or benches were provided instead of upturned bottle cases, the quality of the booze had improved and the gangsters had improved the organisation of their sources of supply. In the long run people refused to be scandalised and those who commit these frauds can never hope to make more than 'a quick buck' before the bubble bursts.

During this immediate post-war decade, America flourished economically and the Dry Set, as they were called, attributed this to Prohibition and proclaimed their beliefs ever louder. Then came 1929, the collapse of confidence in the Wall Street stock market, the Great Depresion, and unprecedented unemployment and poverty among millions of people. President Hoover continued to support Prohibition, although some of his henchmen were beginning to waver. In June 1932, John D. Rockefeller Jr, son of the founder of the Standard Oil Company (Esso), publicly declared:

'I have slowly and reluctantly come to believe that the abolition of the saloon and certain other benefits of Prohibition were far outweighed by the evils incurred. These evils, unless promptly checked, are likely to lead to conditions unspeakably worse than those which prevailed before.'

He was referring not only to the gangsterism with all its brutal violence, but also the bribery and corruption that was rife in all Government and commercial activities. Will Rogers, famed for his wit, commented, 'It's hard to beat a Rockefeller anytime, but when he is right, he's unbeatable.'

The writing was on the wall. Franklin Delano Roosevelt was nominated the Democratic Presidential Candidate and promised the Repeal of Prohibition. He was promptly promised the support of many businessmen, including brewers, distillers and wine growers, who guaranteed the creation of thousands of jobs. This was the talk that pleased millions of Americans and Roosevelt was swept by popular support to an overwhelming victory.

Although he did not take the Oath of Office until 4 March 1933, Congress agreed in the preceding February to submit to the Convention of the individual States, the 21st Amendment to the Constitution, repealing the 18th Amendment - the Act of Prohibition. By December 1933, thirty-six of the forty-eight States had met and signified their approval. Prohibition was finally dead. It was said that all over America, people were singing that jolly song, 'Happy Days are Here Again'.

31

The USA Half Truth

The American wine industry is mainly centred in California. As in Australia, much of the industry is very young. Nevertheless, some of its wines are of excellent quality. The problem is to know what to ask for and what to expect.

Some Californian wines are still named after European wines: Burgundy, Chablis, Claret, Sauternes, Sherry, Rhine, Port and so on, names which were brought over by immigrant settlers. The better class wines bear as a rule the name of the grape, the varietals. These so-called generic wines are not rated among the best produced in the USA and are mostly used for everyday drinking. Their names include Cabernet, Merlot, Shiraz, Zinfandel and others. The Californian winemaker realises, as does his European counterpart, however, that wine made from a single grape variety tends to have a harshness not altogether acceptable. Accordingly, he mixes in a few other varieties of grapes to improve the wine. Fair enough, the quality of the wine is all important. Until new regulations were introduced in line with recent EEC regulations, however, a content of only 51% of a particular variety was sufficient for that name to be given to the wine. The other 49% could have included three or four other varieties. There is no doubt that the wine was improved, but the name was not true.

In American wine's favour, however, 95% of it must be from the year named on the label. Germany, as has been said, need do no better than 85% and France is the same. The revised EEC regulations now include a new clause that wine exported to the USA must contain at least 95% of the stated vintage year.

The contents of Californian sparkling wines vary considerably but they are called 'champagne'. According to the Code of Federal regulations: 'Champagne' is a type of sparkling light wine which derives its effervescence solely from the secondary fermentation of the wine within glass containers of not greater than one gallon capacity, and which possesses the taste, aroma, and other characteristics attributed to champagne as made in the champagne district of France.

Semi-generic designations may be used to designate wines of an origin other than that indicated if there appears in direct conjunction therewith an appropriate appellation of origin disclosing the true place of origin.

Examples of semi-generic names which are also type designations for grape wines are Angelica, Burgundy, Claret, Chablis, Champagne, Chianti, Malaga, Marsala, Madeira, Moselle, Port, Rhine Wine (syn. Hock), Sauterne, Haut Sauterne, Sherry, Tokay.

The new USA regulations regarding generic, semi-generic and geographic description of wines are in preparation at the time of writing. The German government has sent them a brochure in which approximately 1,600 vineyard names, and 1,200 village names are claimed to be geographic designations of origin for their wines. Some of the vineyard names such as Schlossberg, Altenberg, Kirchberg, Moenchberg, Sommerberg, for example, appear in dozens of villages, not only in Germany, but in Austria, Switzerland, Alsace and Luxembourg, and only the combination of a village and a vineyard name is a geographic designation of origin. This brochure is the object of a great scandal, as Germany has not fulfilled her duty to submit the names which her wines bear to the European Commissioner for publication in the official EEC journal, and for English speaking countries she has published an incorrect list! The European Parliament and the International Federation of Wine and Spirit Merchants are fighting this extraordinary chauvinism where Germany tries to cheat the international wine trade.

I expect the USA authorities will give their decision and will help to clear this European muddle and put an end to the German fraud.

32

Alcohol Advertisements in Norway

As soon as any government enacts a law not to the liking of its people, ways will quickly be found to circumvent that law. The prohibition in Norway of advertisements for alcoholic beverages is a typical example. The government officials say that as a result of the law, the consumption of spirits increased much less in 1976 than in previous years. The consumption of moonshine liquor increased substantially, however, and manufacturers of orthodox spirits say that it is well-nigh impossible to launch a new brand.

On 1 March 1977, the ban became absolute and even prohibited the sale of glasses and ashtrays bearing liquor advertisements. In future, they may only be used in authorised bars and restaurants. Liquor manufacturers who supply these items have now ordered more than they ostensibly need, in the positive hope that they will be stolen from the bars and so increase the circulation of the advertisement.

The contents of these advertisements have developed amusingly. Two terriers appeared in one newspaper advertisement; one was black and the other was white. The caption read, 'What do you enjoy most of all, Whitey?' to which the answer was, 'Sorry, but I am not allowed to tell you, Blackie.'

Another advertisement appeared in the form of a strip cartoon. A man is shown stopping outside a State-owned liquor store, looking around cautiously, disappearing inside and then re-appearing with a black and white terrier on a lead. Similar clever cartoons involve white horses, old smugglers, as well as stout, rubicund gentlemen wearing red coats and top hats.

The advertising agencies are clearly enjoying themselves in their ingenuity to promote their clients' interests. Government lawyers must be having an extraordinarily difficult time in trying to decide just how these advertisements are breaking the law.

33

Australia and New Zealand in Trouble

The November 1976 issue of *Grapegrower and Winemaker*, the journal of the Australian wine industry, carried a report by the chairman of the wine judges, Mr B. J. Barry, on the wines exhibited at the 1976 Adelaide Wine Show. Of the 1,550 wines exhibited, 74 received gold medals, 157 silver and 336 bronze; a total of 567, and an indication of some very good wines.

Mr Barry's comments, however, are particularly illuminating after the Salzburg report previously quoted. He stated, 'It was disturbing to see in these classes (dry reds) many examples of first class material spoilt by the presence of H_2S (hydrogen sulphide - the bad egg smell) which good cellar practice can so easily avoid. The disappointing classes were the rosé and to a lesser extent, the dry white table wine, and the main faults here were H_2S, yeastiness, oxidation, amber tints and excess SO_2 (sulphur dioxide - the sulphur smell). The sparkling white wine class had good award wines but the balance was poor, being oxidised, coarse and showing H_2S. The Open dry white table wine full bodied class contained too many light bodied, high acid, young wines and also flavourless wines with basic winemaking faults. The Open burgundies contained many young wines and many of the older wines were starting to deteriorate. It can be said that this class contained some of the best and some of the worst wines in the Show'.

The scandal is that these faulty wines were on sale alongside the other wines of excellent quality. There are at present no Australian wine laws controlling the quality of wines, although these will no doubt be introduced eventually. In the meantime, since most Australians buy their wine either to drink at home or take with them to unlicensed restaurants, they have no means of knowing beforehand whether the wine they have bought is faulty or not. Sadly, it would seem that a fair amount of it is.

On the other hand, it must also be said, that some very fine wines indeed are produced in Australia and are widely available in Britain, as well as at home where there is also a substantial quantity of sound '*vin de consummation courante*' for everyday drinking. Australian wine technology is already advanced and still improving. What is needed in some control to weed out the faulty wines. At present the policy is to encourage the winemaker to do better by comparing his wines with medal winning wines. But demand for wine is increasing every year and some winemakers may not feel the need to take more care with their wines, nor even to exhibit them and so learn from the judges and other experts. The scandal will no doubt persist for a long time to come.

A scandal exists in another form that may eventually end up in the International Court at The Hague. The Australian wine industry owes much of its development and splendid achievement to European winemakers who have, for countless different reasons, left their native land to find a better life in Australia.

Foremost among these are the German Lutherans who emigrated to avoid persecution for their religious beliefs. Many of them settled in the fertile Barossa Valley, just north of Adelaide in South Australia. Indeed, the whole valley has become Teutonic in architecture, social customs and especially in wine nomenclature.

The largest wine co-operative in the Barossa Valley has been named Kaiser Stuhl after the birth place of some of the founders or their fathers, in the wine heartland of Germany. Indeed, the Kaiser Stuhl co-operative and other Barossa wineries have marketed very many wines with German names (and in the familiar fluted bottles, too!) such as Rhinegold, Bernkastel Riesling, Kluges Moselle, Rhine Riesling Moselle, Leo Buring Hock, Liebfraumilch, Liebchenwein, Liebfrauwein, Liebestein, Frauwein, and many others.

Whilst all this was harmless enough at the start and reflected only the German nostalgia for their homeland, the wines have now attained high quality and are being exported not only to Europe, but also to many of Germany's traditional export outlets in the USA, Canada and Asia. Bottle by bottle, there is little to distinguish many Australian wines from German wines, at least in appearance.

It is true that the small print on the label usually includes the words 'Produce of Australia' or the like, but this is usually out of pride of achievement rather than from any legal requirement.

There is a danger that someone unable to understand all the language on the label could easily be misled into buying an Australian wine when he thought he was buying a German wine. This is certainly not due to any fraudulent desire to pass off their wine as German. Germany is far away from Australia and some Australians may not even know that there is a Germany and German wines, as the Germans found out when the Baden Co-operative sued the Barossa Co-operative Winery Ltd (1972). The Australian Registrar of Trademarks considered that Kaiser Stuhl was not a geographic designation and even a German writer (Friese) agreed that Kaiser Stuhl had no such meaning for Australian wine drinkers. I might add that it did not have a geographic connotation for English wine drinkers either. The Germans approached this matter with some chauvinistic feeling and the Australian reaction was quite natural. As the Australian Registrar told the Germans, their German wines are unknown and no confusion was possible.

We have seen so many changes in naming wines and I am sure that the Australians are concerned that nobody should believe their wines to be anything else but Australian. Currently they give more information on their labels than almost any other country. Following a trademark case in the USA, most of the mentioned German names are no longer used - at least for export. I am sure this 'war' can be settled in a friendly way.

The Germans complained about the New Zealand brand of wine called Bernkaizler marketed by the Montana Company. With a great sense of humour the New Zealanders gave up their use of this name and informed their customers as follows: 'In the beginning we simply wanted you to know that here was a fine wine with all the distinctive character of a German-style Riesling. So we called it Bernkaizler. The German Wine Board did not approve. And that, frankly, is embarrassing. For them - because the name Bernkaizler and their own Bernkastel are too close for comfort. For us - because we are proud of our wine and don't need to copy anyone. So we've decided to change the name - to bring Bernkaizler right back home to New Zealand, where it belongs. With a new name derived from a peak overlooking the Montana Marlborough vineyards and the name of the Estate in the Gisbourne area. The Riesling you know and love is now called Benmorven Riesling.'

Well done New Zealand! And the Stabilierungsfonds? We have met before with their false propaganda. It was interesting to read an article by their executive director, Dr Michel, as follows: 'What perhaps 50 or 100 years ago acted as a memory of the fatherland has now become a systematical desire for profit when the naming of imitations are selected according to the image of the original. This goes so far that a wine company in New Zealand used the following happy slogan: "We admit that French Champagne is as good as ours".'

Poor Dr Michel – just a *little* sense of humour, please!

34

Know Your Wine

By now it will be clearly seen that fraud, deceit, doctoring, dishonesty, adulteration, labelling with intent to mislead, false descriptions and countless other wrong doings, are part and parcel of the wine trade as a whole from the grower to the retailer, perpetrated by some black sheep within it.

The fact must be recognised that to millions of people wine is just a saleable commodity. For them it has no mystique, no enchantment. If circumstances were different for those people, they might be making and selling soap or sausages. People produce and market wine solely to make from it the best living that they can for themselves and their families. This is not an unethical attitude in itself and happily there are many proud and honest craftsmen in the growing, the making and the marketing of wine. The difficulty is in finding them. As in so many matters, it is the corrupt and the greedy ones who engender the bad reputation from which the innocent must suffer.

It would be helpful if one could always follow the recommendations of wine journalists. But they, too, are fallible, and the ability to write does not necessarily imply an ability to evaluate wine. They can be deceived as much as we can, and they have to be careful in their criticisms that they do not offend against the laws of libel. They have to remain on good terms with growers, brokers and shippers in order to keep in touch with their subject.

BUYING WINES

The greatest aid to the buying of good wine and to avoid becoming the victim of a wine scandal is surely a reputable wine merchant. The managers of most multiple-branch wine shops receive some formal training before assuming their responsibility. Some are even sent to the wine countries to learn about the wines at first hand. In any case they handle a wide range of wines and with the passing years usually manage to acquire a worthwhile knowledge of their subject. Some privately owned businesses are often directed by an oenophile who often buys wines of his own choice for sale in his shops. It is well worthwhile taking the trouble to find such a person and making his acquaintance. If you ask for and accept his advice and then tell him in a kindly and constructive manner what you thought of the wines, you will soon find that a friendship springs up between you. There is a very special bond between wine lovers all over the world and you should take advantage of it.

Very likely your wine merchant will hold some tastings from time to time. Do attend if you can. Mark your card with your likes and dislikes and buy some stock of those that you liked. Apart from having wines that you will enjoy, you will save yourself the cost of buying those that you did not like but might otherwise have bought on their reputation or their label. A label may tell you from where the wine comes, who made it and when, but it cannot tell you anything about the quality of the wine in the bottle. Only your own palate can provide this important piece of information.

There is an old saying in the wine trade, 'Buy on an apple and sell on a cheese.' With the sharp taste of an apple in your mouth, the wine has to be good to taste good. On the other hand, the fatty taste of cheese smothers any acid taste of a wine. Be warned then. Buy when you feel well and free from worry. Look, smell and taste before you buy. Make your choice slowly and free from pressures.

When you purchase a quantity of wine from a merchant, care should be taken about leaving it with him while it matures in his cellars. It is imperative to insist that your parcel of wine be labelled with your name and address as well as the number, or a copy, of your receipt. Preferably the bottles should all be in one sealed container and set aside well marked from others and from the merchant's stock. There should never be any doubt, then, as to which wines belong to you. The reasons for this are twofold: if your wines are not so marked, an uninformed assistant may unwittingly sell your wines to some other person. Furthermore, should the firm become bankrupt, there is no certain indication that these wines are specifically yours and not part of the merchant's stock which can be liquidated. A receipt by itself may be regarded merely as a payment in advance, in which case all the money would be lost. Only if your wines are clearly identifiable as yours are they safe. There have been numerous court cases where failure to take adequate precautions has resulted in the loss of the wine. But it must also be mentioned that in such a court case, the bankers who had first rights on the stock, generously allowed it to be distributed among the buyers who believed that their stock was safely lying in the vendor's cellars.

STORING WINE

Should you decide to take the wines home with you, store the bottles on their sides in a place where the temperature remains as even as possible. Although 12°C (54°F) is recommended, wine is reasonably tolerant of a temperature a little higher or even a little lower. What it does not like are sudden and substantial fluctuations. The lower the temperature the longer you have to wait until the wine is fully mature.

DRINKING WINE

Store the wine *sensibly* and serve it well. This means choosing a suitable wine for the purpose, ensuring that it is adequately mature, that it is at the right temperature for drinking, and is served in a proper glass and accompanied by some complementary food. It tastes even better when drunk in the company of like-minded friends.

Should a wine disappoint you after all, never throw it away; no matter whether it be the full bottle that you do not like or the remnant of one that you left undrunk so long that it became undrinkable. There is no better spice in the world than wine - wine in the casserole, the pie, the stew, the gravy - never waste wine!

I very much hope that no reader of my book will be put off drinking wine because of the fiddles and scandals that I have described. If you select a reputable wine supplier and buy your wine to suit your own palate and liking you will not become the victim of some wine fraud. Drink wine for your health, for your enjoyment and forget all about wine scandals.

Appendix

EEC Reg 337 79: Oenological Practices, Annex III

(Reproduced from the *Official Journal of the European Communities*)

1. Oenological practices and processes which may be applied to fresh grapes, grape must, partially fermented grape must, concentrated grape must and new wine still in fermentation:
   (a) aeration;
   (b) thermal treatment;
   (c) centrifuging and filtration, with or without an added inert filtering agent on condition that no undesirable residue is left in the product so treated;
   (d) use of carbon dioxide or nitrogen either alone or combined, in order to create an inert atmosphere and to treat the product shielded from the air;
   (e) use of yeasts for wine production;
   (f) addition of diammonium phosphate or ammonium sulphate up to 0.3 g/l respectively and of thiamin hydrochloride up to 0.6 mg/l expressed as thiamin to encourage the growth of yeasts;
   (g) use of sulphur dioxide, potassium bisulphite or potassium metabisulphite which may also be called potassium disulphite or potassium pyrosulphite;
   (h) elimination of sulphur dioxide by physical processes;
   (i) treatment of white must and new white wines still in fermentation with charcoal for oenological use, up to a maximum of 100 g of dry product per hectolitre;
   (j) clarification by means of one or more of the following substances for oenological use:
   - edible gelatines,
   - isinglass,
   - casein and potassium caseinate,
   - animal albumin (egg albumin and dried blood powder),
   - bentonite,
   - silicon dioxide as a gel or colloidal solution,

- kaolin,
- tannin,
- pectinolytic enzymes;

(k) use of sorbic acid or potassium sorbate;

(l) use of tartaric acid for acidification purposes under the conditions laid down in Articles 34 and 36;

(m) use of one of the following substances for deacidification purposes under the conditions laid down in Articles 34 and 36:
- neutral potassium tartrate,
- potassium bicarbonate,
- calcium carbonate, which may contain small quantities of the double calcium salt of L (+) tartaric and L (−) malic acids.

2. Processes and oenological practices which may be applied to partially fermented grape must intended for direct human consumption in its natural state, wine suitable for producing table wine, table wine, sparkling wine and quality wines psr:

(a) use in dry wines, and in quantities not exceeding 5% of fresh lees which are sound and undiluted and contain yeasts resulting from the recent vinification of dry wines;

(b) aeration;

(c) thermal treatment;

(d) centrifuging and filtration, with or without an added inert filtering agent on condition that no undesirable residue is left in the product so treated;

(e) use of carbon dioxide or nitrogen, either alone or combined, to create an inert atmosphere and to treat the wine shielded from the air. The carbon dioxide content of wine preserved or treated in this way may not exceed 2 g/l;

(f) addition of carbon dioxide, provided that the carbon dioxide content of wine so treated does not exceed 2 g/l;

(g) use, as laid down in Community rules, of sulphur dioxide, potassium bisulphite or potassium metabisulphite, which may also be called potassium disulphite or potassium pyrosulphite;

(h) addition of sorbic acid or potassium sorbate provided that the final sorbic acid content of the treated product on its release to the market for direct human consumption does not exceed 200 mg/l;

(i) addition of up to 150 mg/l of L-ascorbic acid;

(j) addition of citric acid for wine stabilisation purposes, provided that the final content in the treated wine does not exceed 1 g/l;

(k) use of tartaric acid for acidification purposes under the conditions referred to in Articles 34 and 36;

(l) use of one of the following substances for deacidification purposes under the conditions referred to in Articles 34 and 36:
- neutral potassium tartrate,
- potassium bicarbonate,

- calcium carbonate which may contain small quantities of double salt of calcium of L (+) tartaric and L (−) malic acids;

(m) clarification by means of one or more of the following substances for oenological use:
- edible gelatines,
- isinglass,
- casein and potassium caseinate,
- animal albumin (egg albumin and dried blood powder),
- bentonite,
- silicon dioxide as a gel or colloidal solution,
- kaolin;

(n) addition of tannin;

(o) treatment of white wines with charcoal for oenological use, up to 100 g of dry product per hl;

(p) treatment, under conditions to be laid down:
- of white wines and rosé wines, with potassium ferrocyanide,
- of red wines, with potassium ferrocyanide or with calcium phitate, in accordance with the sixth subparagraph of Article 46 (3);

(q) addition of up to 100 mg/l of metatartaric acid;

(r) use of acacia;

(s) use of DL tartaric acid, under conditions to be laid down, for precipitating excess calcium;

(t) use of oenocyanin under the conditions laid down in Article 46 (3);

(u) use of sodium-based cation exchange resins under the conditions referred to in Article 46 (3);

(v) use of discs of pure paraffin impregnated with allyl isothiocyanate to create a sterile atmosphere, solely in Member States where it is traditional and in so far as it is not forbidden by national law, provided that they are used only in containers holding more than 20 litres and that there is no trace of allyl isothiocyanate in the wine;

(w) treatment by silver chloride under the conditions laid down in Article 46 (3) provided that the silver content of the product so treated is not more than 0·1 mg/l;

(x) treatment by up to 20 mg/l of copper sulphate under the conditions laid down in Article 46 (3) and provided that the copper content of the product so treated is not more than 1 mg/l.

24 LIEBFRAUMILCH
77 MOSEL-SAAR-RUWER